Farid E. Ahmed, PhD
Editor

Testing of Genetically Modified Organisms in Foods

Pre-publication
REVIEWS,
COMMENTARIES,
EVALUATIONS . . .

"This book brings together, in one technically detailed volume, the collective experience of world-class experts in the field of GM testing. The result is an informed and balanced work describing existing and potential methods that can be used to detect the presence of GM in foods. This book is an invaluable resource for individuals in the food industry who carry out due diligence exercises on their products, regulators who wish to deepen their background knowledge on the effectiveness of tests, and of course, laboratory personnel of all levels who are involved in carrying out tests.

The chapters on molecular and immunological methods are authoritative and detailed in their scope and application. They are ably supported for the less-experienced reader by the basic instruction provided in the adjunct chapters. The instructive chapters describing the production of reference materials and sampling serve to provide a practical dimension, and remind the reader that a very real issue is being addressed. The book describes the multifaceted industry that has grown around GM testing, since transgenic materials were first approved for food use. This comprehensive elucidation of the complexity and detail required to ensure that sampling and testing are effectively carried out as required by legislators will come as a surprise to many, and as a reassurance to others. I am looking forward to having a first edition copy of this important reference work on my bookshelf."

Sarah Oehlschlager, PhD
Team Leader, GM Testing,
Central Science Laboratory, United Kingdom

More pre-publication
REVIEWS, COMMENTARIES, EVALUATIONS . . .

"**A**hmed masterfully weaves the reader through the complex, politically charged, and evolving field of food labeling. The authors provide the reader with balanced coverage of the complex issues associated with the risks and benefits of agricultural biotechnology, cover the regulations currently supporting food and feed labeling requirements, and provide a comprehensive review of the various analytical methods used to detect DNA and protein in food. This is a super source and a great desk reference for anyone considering developing sampling strategies and methods to support food and feed labeling requirements, for those new to the field of agricultural biotechnology, or for seasoned veterans struggling to keep up with the fast pace of this developing field."

Glen Rogan, MSc
Regulatory Affairs Manager,
Monsanto Company

"**T**here is an increasingly overwhelming consensus among scientists and physicians that the current process for assessing the safety of GM crops as foods is adequate and that foods from crops that survive the regulatory process are as safe as conventionally produced foods. However, the public is generally not convinced, and has successfully convinced food companies and politicians that GM crops must be traced, with foods tested for GM content and labeled.

This book outlines in considerable detail methods that can be used to meet these labeling requirements, and is very balanced in its considerations of sampling methodology, reference material and standards, protein and DNA-based methods, and the limitations of near-infrared spectroscopic methods."

Rick Roush, PhD
Director, Statewide IPM Program,
University of California

Food Products Press®
An Imprint of The Haworth Press, Inc.
New York • London • Oxford

Testing of Genetically Modified Organisms in Foods

FOOD PRODUCTS PRESS®
Crop Science
Amarjit S. Basra, PhD
Senior Editor

Testing of Genetically Modified Organisms in Foods

Farid E. Ahmed, PhD
Editor

CRC Press
Taylor & Francis Group
Boca Raton London New York

CRC Press is an imprint of the
Taylor & Francis Group, an **informa** business

CONTENTS

ABOUT THE EDITOR

Farid E. Ahmed, PhD, is Adjunct Professor of Radiation Oncology at the Leo W. Jenkins Cancer Center at the Brody School of Medicine, East Carolina University. He is also Director of GEM Tox Consultants, Inc., which is a consulting firm specializing in food safety, toxicology, and biotechnology. He has authored over 130 peer-reviewed scientific publications, serves on the editorial boards of several journals in toxicology, environmental sciences, and molecular pathology, and is a member of various scientific societies and educational organizations.

CONTRIBUTORS

Robert P. Cogdill, MS, Research Associate, AGROMETRIX/ Camagref, Montpellier Cedex 1, France.

Philippe Corbisier, PhD, Scientific Officer, European Commission, Joint Research Centre (EC-JRC), Reference Materials and Measurements Units, Institute for Reference Materials and Measurements (IRMM), Geel, Belgium.

John Fagan, PhD, CEO and Chief Scientific Officer, Genetic ID, Fairfield, Iowa.

Larry D. Freese, MS, Mathematical Statistician, United States Department of Agriculture, Federal Grain Inspection Service, Packers and Stockyards Administration, Technical Service Division, Kansas City, Missouri.

Claudia Paoletti, PhD, Scientific Officer, European Commission, DG JRC, Institute for Health and Consumer Protection, Biotechnology and GMO Unit, Ispra, Italy.

Sylvie A. Roussel, PhD, CEO, AGROMETRIX/Cemagref, Montpellier Cedex 1, France.

Heinz Schimmel, PhD, Scientific Officer, European Commission, Joint Research Centre (EC-JRC), Reference Materials and Measurements Units, Institute for Reference Materials and Measurements (IRMM), Geel, Belgium.

Anthony M. Shelton, PhD, Professor, Cornell University/New York State Agricultural Experimental Station, Department of Entomology, Geneva, New York.

James W. Stave, PhD, Vice President, Research and Development, Strategic Diagnostics, Inc., Newark, Delaware.

Stefanie Trapmann, PhD, Scientific Officer, European Commission, Joint Research Centre (EC-JRC), Reference Materials and Measurements Units, Institute for Reference Materials and Measurements (IRMM), Geel, Belgium.

Preface

This book addresses an important and emotionally charged issue in the food safety arena: eating food modified by organisms whose genetic material has been intentionally tampered with to introduce traits not normally present in the original food. This subject has attracted worldwide interest and is highly controversial in most countries due to differences in opinion as to the potential risks to human health and/or the environment associated with this procedure. These organisms, which are often called recombinant DNA, genetically modified, genetically engineered, bioengineered, or biotechnology manipulated organisms, are generally referred to as GMOs (genetically modified organisms), and the food is known as genetically modified or GM food. GMOs produce modified genetic material (i.e., DNA) and novel protein(s) in consumed food.

I have striven to produce a book (the first of its kind that is not produced as a result of a scientific meeting or proceedings of a symposium) that incorporates the latest developments in testing GMOs, and have tried to address the risks—whether real or imagined—that result from their presence in food. In addition, I have tried to present this information in a balanced way that incorporates all viewpoints and concerns raised by the various groups interested in the subject, including those from in the European Union, United States, and other countries.

Contributors to this book were selected from Europe and the United States on the basis of their recognized expertise in certain topics as demonstrated by their publications in the open reviewed literature and their presentations at public and scientific meetings. Thus, this book represents a state-of-the-art collection on the topic of genetically modified food and is a primary source of reference—worldwide—for college students enrolled in food safety courses, food technologists and scientists involved in testing GMOs (or attempting to establish testing methods in their respective laboratories), food handlers and suppliers of food ingredients, quality control/assurance staff, analytical chemists and molecular biologists working in the food sector, scientists and administrators within the regulatory framework at all levels, and those at the sharp end of implementation of standards or involved in the provision of advice to the public, industry, or governments.

Contributors range from food and environmental scientists and analysts in academics or public research institutes who are dealing with the day-to-day issues raised by the monitoring and surveillance of these organisms to staff within the regulatory authorities who are continuously assessing the information and advice from all sources, to scientists in genetic testing facilities who are often involved in advising industry, governments, consumers, and consumer protection organizations on testing regimens worldwide.

Development of this book took place during ongoing legislation and regulatory debates in the European Union and the United States. The contents cover the areas of risks and benefits of GMOs in the food supply and the environment, sampling concepts and plans in both the European Union and the United States, reference material and standards, protein-based methods and DNA-based methods of testing, near-infrared spectroscopy, and other methods that may be of academic/research value or applicable to field testing (e.g., chromatographic, spectrometric, and nuclear magnetic resonance-based methods; biosensors; DNA chips and microarray technology; and new developments in proteomics research).

I hope that the overall combination of contributors and content, which brings together experts from opposite scientific and regulatory cultures candidly expressing their own views and beliefs, is timely and provides deeper insight into the issue at hand. Because a learning process is facilitated by considering the experiences of others with, perhaps, a different set of objectives, priorities, and beliefs, it is also hoped that the broad exchange of ideas lying within these chapters will help guide the reader to the real issues that impact consumers of food containing GMOs.

I gratefully acknowledge the amicable working relationships that developed between the authors, the publisher, and myself. Despite a variety of difficulties (e.g., ill health, work priorities, job pressures, and personal problems), chapters were diligently produced and my prodding queries and suggestions were patiently dealt with. I am also indebted to my family—my wife, Karen, and my sons, Khaled and Salah—for enduring my long work hours to produce this book in a timely manner, and to my colleague Roberta Johnke for her editorial help.

Chapter 1

Risks and Benefits of Agricultural Biotechnology

Anthony M. Shelton

THE DEVELOPMENT OF BIOTECHNOLOGY

Over the past 10,000 years, humans became food producers by increasing the numbers of domesticated plants and animals and modifying them through selective breeding. Only in the past century have breeders used techniques to create crosses that would not have been viable in nature, and this has been accomplished through modern biotechnology. The term *biotechnology* has evolved over time to take on new meanings, and it has become one of the most used and abused words in modern biology (Brown et al., 1987). In a broad sense it can be defined as using living organisms or their products for commercial purposes, and according to this definition, biotechnology has been around since the beginning of recorded history in animal and plant breeding, brewing beverages, and baking bread. A more modern definition focuses on the deliberate manipulation of an organism's genes using a set of techniques of modern biology that employs living organisms (or parts of organisms) to make or modify products, improve plants or animals, and/or develop microorganisms for specific uses. Genetic engineering is one form of biotechnology that involves copying a gene from one living organism (plant, animal, or microorganism) and adding it to another organism. Today's breeders may define a genetically engineered organism as a living thing that has been im-

I wish to thank the many scientists whose work has helped elucidate the many complex issues surrounding biotechnology, members of the Cornell Agricultural Biotechnology Committee for their fruitful discussions, and H. L. Collins for her help in preparing this chapter.

proved using genetic engineering techniques in which only a small piece of one organism's genetic material (DNA) is inserted into another organism. Products of genetic engineering are often commonly referred to as "genetically engineered organisms," "GE products," "genetically modified organisms," or "GMOs."

The techniques employed in biotechnology, especially those used in genetic engineering, allow plants and animals, as well as many nonagricultural products such as medicines, to be developed in ways that were not thought possible only a few decades ago. The results have led to what is considered the third technological revolution, following the industrial and computer revolutions (Abelson, 1998). Discoveries in disciplines from physics to genetics have built on one another to progressively create today's biotechnology. The work of James Watson, Francis Crick, Maurice Wilkins, and Rosalind Franklin in the early 1950s led to an understanding of the structure of DNA as the carrier of genes, the "inheritance factors" noted by the Austrian monk Gregor Mendel, who is often considered the founder of genetics. Others have followed in the long line of scientists who were able to use basic biology to understand how genes function in living organisms. IN the late 1960s, Paul Berg delineated the key steps by which DNA produces proteins, and this became an important step in the future of recombinant DNA techniques and genetic engineering. Scientists soon came to realize that they could take specific segments of DNA that carried information for specific traits (genes) and move them into other organisms. In 1972, the collaboration of Herbert Boyer and Stanley Cohen resulted in the first isolation and transfer of a gene from one organism to a single-celled bacterium that would express the gene and manufacture a protein. Their discoveries led to the first direct use of biotechnology—the production of synthetic insulin to treat diabetes—and the start of what is often called modern biotechnology (Kelves, 2001).

In the ongoing dialogue about biotechnology, it is important to understand how it is similar and how it differs from more traditional aspects of breeding. For example, traditional plant breeding relies on artificial crosses in which pollen from one species is transferred to another sexually compatible plant. The purpose of the cross is to bring desirable traits, such as pest resistance, increased yield, or enhanced taste, from two or more parents into a new plant. Plant breeding depends on the existence of genetic variation and desirable traits.

However, desirable characteristics are often present in wild relatives not sexually compatible with the parent plant, so other means of transferring the genetic material are needed. Since the 1930s, plant breeders have used various techniques to allow them to create novel crosses and new varieties of plants. Some of these techniques fall under the broad classification of biotechnology but do not include genetic engineering (i.e., the cloning of a gene and moving it to another organism). An example of a technique that does not include genetic engineering is *embryo rescue*, in which the offspring of the cross would not survive without special help provided in the laboratory.

Beginning in the 1950s, plant breeders also used methods of creating artificial variation in an organism by using radiation and chemicals that randomly caused mutations or changes in the genes of the plant (in some aspects this is similar to what occurs naturally through exposure to solar radiation or naturally occurring chemicals). Plants were then assessed to determine whether the genes were changed and whether the change(s) gave the plant some beneficial trait, such as disease resistance. If the plant was "improved" by this technique, it was tested further for any negative effects caused by the treatment. Many of the common food crops we use today were developed with agricultural biotechnology techniques such as radiation, chemical breeding, and embryo rescue. Some have been developed with the more recent advances in genetic engineering. In the late 1970s and early 1980s, Chilton and colleagues produced the first transgenic plant. Parallel to the efforts to create plants with novel traits through genetic engineering, research was being conducted to develop other agriculturally related products through recombinant DNA (rDNA) technology. Today's milk is commonly obtained from cows treated with a genetically engineered hormone called bovine somatotropin (bST), which is used to increase milk production.

Many scientists consider genetic engineering to be a continuation of traditional breeding. In the case of plant breeding, however, two major differences exist between traditional plant breeding (which also includes many techniques involving agricultural biotechnology, as noted earlier) and genetic engineering. The first is the amount of genetic material involved. When two parental plant lines are crossed using traditional breeding methods, the new plant obtains half the genetic makeup of each parent, and the desirable gene may be accompanied by many undesirable genes from that same parent. To remove the undesirable genes, continued breeding is required. In the case of genetic engineering, only the few genes that are specifically desired are moved into the new plant, and it is not necessary to remove all of the undesirable genes. However, as in traditional plant breeding, it is important to assess how that desired gene is expressed.

A second difference between traditional breeding and genetic engineering is the source of genetic material used. Traditional breeding relies on closely related plant species. In genetic engineering, at least theoretically, a gene from any living organism may be moved into another living organism. For example, this process permits scientists to move genes from a bacterium, such as *Bacillus thuringiensis* (Bt), into a plant to protect it from insect attack. Prior to Bt-insect-protected plants, the development of insect-resistant plants relied on traditional breeding to produce characteristics in plants that would be deleterious to the insects—primarily increased levels of chemicals that would affect the development or behavior of the insect. Genetic engineering allows the possibility of a multitude of other chemicals to be expressed in plants, even chemicals from organisms distantly related to the plant into which it will be incorporated. One of the reasons genes have been moved so successfully between seemingly different organisms such as plants and bacteria is that all living organisms share the same bases that make up DNA and the synthesis of proteins and other basic life functions.

Hundreds of products containing ingredients derived through genetic engineering are sold in marketplaces throughout the world. These include medicines (e.g., insulin), fuels (e.g., ethanol), biomaterials (e.g., detergents), and many food products (e.g., cheeses). Although genetically engineered products were first introduced into food products in 1990—chymosin, an enzyme used for cheese production; and a yeast for baking—they were not widely publicized and did not elicit much public reaction. Most of the present controversy about agricultural biotechnology has revolved around the first wave of genetically engineered plants, i.e., those used to manage insects, weeds, or diseases (Shelton et al. 2002). A second wave of genetically engineered products is waiting in the wings and will be far more varied in scope, including fish that grow faster, animals that produce less manure, plants that decontaminate soils, and plants that produce medicines (Abelson, 1998).

BEGINNING OF THE CONTROVERSY
ABOUT BIOTECHNOLOGY

Following the first recombinant DNA experiments by Boyer and Cohen, 11 scientists, including Boyer, Cohen, Berg, and Watson, published a letter in *Science* calling for a moratorium on most recombinant DNA work pending a review of its potential hazards (Berg et al., 1974). Under the sponsorship of the National Academy of Sciences of the United States, 140 people attended the Asilomar Conference to discuss the potential hazards of this new science (Berg et al., 1975). Out of these discussions came a set of recommendations adopted by the National Institutes of Health (NIH) to govern federally sponsored work on rDNA. The guidelines were intended to protect the health and safety of the public by providing special containment facilities or by engineering organisms so they could not survive (Kelves, 2001). Despite these guidelines, recombinant DNA work became highly controversial not only within the scientific community but also in the public sector. When Harvard University decided to create an rDNA containment facility at it's Cambridge, Massachusetts, campus in the 1970s, the mayor of Cambridge proposed a ban on all recombinant DNA work. Although the mayor's proposal was turned down, a panel of laypeople proposed that such work could proceed but under even stronger safeguards than those proposed by NIH. Soon after, the U.S. Congress joined state and local authorities to develop new and tougher regulations for recombinant DNA.

Many scientists, including some who proposed the original moratorium, soon came to believe that they had overstated the potential risks of biotechnology at the expense of limiting its potential benefits (Kelves, 2001). As scientists gained more experience with rDNA and were better able to judge any associated risks, NIH guidelines gradually changed and allowed more experimentation. However, as rDNA experiments moved from the laboratory into agricultural fields, controversy began anew. In 1983 a proposal to field-test a genetically altered bacterium (Ice-Minus), sprayed onto plants to reduce the risk of freezing, was delayed for four years through a series of legal challenges (Carpenter, 2001). This was the beginning of a series of controversies about risk assessment in modern agricultural biotechnology that generally focused on issues of food safety and the environment.

RISK AND BENEFIT ANALYSIS

As with any technology, uncertainties exist in our knowledge about the risks and benefits of agricultural biotechnology. Some degree of hazard is associated with every technology and all activities of our daily lives, but it is important to assess the likelihood and consequences of these risks and compare them to the potential and actual benefits. Risk and benefit analysis should be an ongoing process for all technologies, even ones that have been in existence for decades. Assessment of benefits is a process perhaps less controversial than its flip side—assessment of risks. Risk assessment involves four steps: hazard identification, dose-response evaluation, exposure assessment, and risk characterization (NAS, 2000). Hazard identification relates to a particular item causing a documented adverse effect. Dose-response evaluation involves determining the relationship between the magnitude of exposure and probability of the adverse effect. Exposure assessment can be defined as the set of circumstances that influence the extent of exposure. Risk characterization is a "quantitative measurement of the probability of adverse effects under defined conditions of exposure" (NAS, 2000). Although scientists may assert that such risk assessments, despite problems of variability and extrapolation, are needed for science-based decisions, the theory of risk assessment is not easily explained to the general public (Shelton et al., 2002). Although no agricultural management practice is without risk, the public's attention has been focused more on the risks of biotechnology than on the risks of the alternatives.

There are different philosophies concerning any potentially harmful aspects of producing GM crops or consuming products from them: (1) risk assessment, favored in the United States, which tries to balance risk with public health and benefits; and (2) the precautionary principle, used in some international treaties and increasingly in Europe, which places more emphasis on avoiding any potential risk and less emphasis on assessing any potential benefits. The precautionary principle is often invoked in cases in which limitations to addressing risk assessment exist. A common definition of the precautionary principle is the so-called Wingspread Declaration: "When an activity raises threats of harm to human health or the environment, precautionary measures should be taken, even if some of the cause-and-effect relationships are not established scientifically" (Goklany,

2002). If a new technology is "risk-neutral," then the choice of whether to use it is easy. However, most policy options depend on an assessment of the risks of one technology compared to the risks of another. Goklany (2002) suggests that to ensure a policy is actually precautionary—i.e., it reduces net risk—one should compare the risks of adopting the policy against the risks of not adopting it. This inevitably results in a risk-risk analysis. Thus, despite claims that risk analysis differs from the precautionary principle, "the latter logically ends up with risk-risk analysis" (Goklany, 2002). Furthermore, if the argument is made that the precautionary principle is a risk-risk assessment, then there should be a framework in which it can operate, assuming it is not a risk-neutral event. In this framework, the risks of using or not using a technology should be ranked and compared based on "their nature, severity, magnitude, certainty, immediacy, irreversibility and other characteristics" (Goklany, 2002). Differences in perception about whether to use a strict precautionary principle or a more formal risk-benefit analysis have profound consequences in international regulations and trade of GM crops and products derived from them.

THE PRESENT SCOPE OF AGRICULTURAL BIOTECHNOLOGY PRODUCTS

From their first commercialization in 1996, when they were grown worldwide on 1.7 million hectares (ha), GM crops have increased more than 30-fold to 52.6 million ha in 2001 (James, 2001b). The estimated number of farmers producing GM crops worldwide is 5.5 million, and the largest adopters of GM crops in 2001 were the United States (68 percent of the global total), followed by Argentina (22 percent), Canada (6 percent), and China (3 percent). In addition, small amounts of GM crops were also produced in South Africa, Australia, India, Mexico, Bulgaria, Uruguay, Romania, Spain, Indonesia, and Germany. From 2000 to 2001, GM crops increased by 19 percent, despite the multifaceted international debate about their use. The main GM crops were soybeans (63 percent of global area), followed by corn (19 percent), cotton (13 percent), and canola (5 percent). The main GM traits produced in these crops are herbicide tolerance (77 percent of total GM area), insect resistance (15 percent), and herbicide-tolerance plus insect resistance (8 percent). Of the total worldwide crop grown in 2001, 46 percent of the soybean crop was GM, 20 percent of cotton, 11 percent of canola, and 7 percent of maize. Adoption of this new technology, as with most other technologies, has been fastest in the industrialized countries, but the proportion of transgenic crops grown in developing countries has increased consistently from 14 percent in 1997, to 16 percent in 1998, to 18 percent in 1999, and to 24 percent in 2000 (James, 2000).

Even with the rapid development of GM crops, it is important to remember that pest-management crops should be deployed as part of an overall crop-management strategy that relies on the use of cultural and biological controls that are compatible with GM technologies. Agriculture is a complex biological system that requires managing pests within a holistic framework for a more sustainable system. As with any technology, the limitations of biotechnology crops must be kept in mind so they can be used in the most sustainable fashion possible.

Herbicide Tolerance

Glyphosate, commonly known as Roundup, is a broad-spectrum herbicide that controls weeds by inhibiting the enzyme 5-enolpyruvylshikimate-3-phosphate synthetase (EPSPS) that catalyzes the synthesis of amino acids essential for the survival of plants. EPSPS is found in all plants, fungi, and bacteria but is absent in animals (Padgette et al., 1996). Glyphosate binds with EPSPS and inhibits its activity to produce aromatic amino acids, which leads to cell death. A glyphosate-tolerant EPSPS from the soil bacterium *Agrobacterium* sp. strain CP4 was isolated and then introduced into the genome of a soybean cultivar so that when the cultivar and any surrounding weeds were treated with glyphosate, the weeds but not the soybeans would die. All currently commercialized glyphosate crops (Roundup Ready), including corn, cotton, canola, and soybean, contain a tolerant EPSPS gene obtained from one or two sources (Carpenter et al., 2002). One cultivar of GM corn does not use the CP4 strain but instead uses the EPSPS gene altered by chemical mutagenesis.

Herbicide-tolerant (HT) soybean, grown on 33.3 million ha in 2001, is the most extensively planted biotech crop (James, 2001b). Soybeans (both HT and non-HT) are grown on 72 million ha worldwide, with the United States producing 41 percent, Brazil 18 percent, Argentina 13 percent, China 11 percent, and India 5 percent (USDA NASS, 2001). Herbicide tolerance is also used on 2.7 million ha of canola, 2.5 million ha of cotton, and 2.1 million ha of maize. Additional acreage of HT cotton and maize are also grown with "stacked genes," i.e., genes for different functions that are combined in the same plant, for insect and herbicide resistance.

Insect-Resistant Crops (Bt Crops)

The common soil bacterium *Bacillus thuringiensis* (Bt) has been commercially used for more than 50 years as an insecticide spray. The insecticidal activity of commercially used Bt comes from endo-toxins included in crystals formed during sporulation. The crystals of different strains of most Bts contain varying combinations of insecti-cidal crystal proteins (ICPs), and different ICPs are toxic to different groups of insects (Shelton et al., 2002). When ingested, the sporu-lated Bt cells are solubilized in the alkaline midgut of the insect, and protein toxin fragments then bind to specific molecular receptors on the midguts of susceptible insects. Pores are created in the insect gut, causing an imbalance in osmotic pressure, and the insect stops feed-ing and starves to death (Gill et al., 1992). More than 100 Bt toxin genes have been cloned and sequenced, providing an array of pro-teins that can be expressed in plants or in foliar applications of Bt products (Frutos et al., 1999). Insecticidal products containing sub-species of *B. thuringiensis* were first commercialized in France in the late 1930s, but even in 1999 the total sales of Bt products constituted less than 2 percent of the total value of all insecticides (Shelton et al., 2002). Bt, which had limited use as a foliar insecticide, become a ma-jor insecticide when genes that produce Bt toxins were engineered into major crops. By the late 1980s, tobacco, tomato, corn, potato, and cotton had been transformed to express Bt toxins.

Insects targeted for control through the production of Cry1Ab, Cry1Ac, and Cry9C proteins by Bt plants are primarily the immature stages of Lepidoptera (caterpillars), although one product has been developed for control of the Colorado potato beetle using Cry3A. In 2001 on a worldwide basis Bt plants were grown on 12.0 million ha with 4.2 million ha of that total being plants with stacked genes for herbicide resistance (James, 2001b). The area of Bt maize grown is double that of Bt cotton. In addition to maize and cotton, Bt potatoes have been commercialized, but they failed to capture more than 4 percent of the market and are no longer sold in North America. A primary reason for the market failure of Bt potatoes was the introduction of a new class of insecticides that is effective against the potato beetle as well as aphids.

Efforts are underway to commercialize maize with a *Cry3Bb* gene or a binary toxin genetic system for control of the corn rootworm *(Diabrotica)* complex. Corn rootworm annually causes losses and control costs estimated at $1 billion in the United States (Metcalf and Metcalf, 1993) and is heavily treated with insecticides. Other Bt crops under development are canola/rapeseed, tobacco, tomato, apples, soybeans, peanuts, broccoli, and cabbage (Shelton et al., 2002).

Virus Resistance

Conventional breeding has developed a number of virus-resistant cultivars of important crops, including potatoes, wheat, corn, and beans. However, virus resistance genes have not been identified in wild relatives of many crops, so genetic engineering has been employed. In the 1980s scientists demonstrated in a plant that the expression of a coat protein (CP) gene from a virus could confer resistance to that virus when it attempted to infect the plant (Tepfer, 2002). Since this technology was first developed, a large number of virus-resistant transgenic plants have been developed using "pathogen-derived resistance" techniques. Squash and papaya have been engineered to resist infection by some common viruses, and are approved for sale in the United States. Two virus-resistant varieties of papaya are grown in Hawaii and are awaiting deregulation for the Japanese market.

REGULATIONS ON GROWING BIOTECHNOLOGY CROPS

In 2000, biotech crops were produced in a total of 13 countries, and each country had its own regulatory system. In addition to the regulations pertaining to the production of biotech crops, each country may have additional regulations on the importation of biotech crops or on whether products derived from biotech crops must be labeled. Clearly no global standards exist presently for biotech crops, and regulatory agencies are being challenged by the evolving science of biotechnology. In the European Union (EU), member countries have not agreed on a standard policy, although some countries, such as France, Germany, and Spain, do grow some biotech crops. A new directive by the European Union became effective in fall 2002, requiring more environmental monitoring, as well as labeling and tracking of biotech products through all stages of the food chain. In the past, some countries have not followed such directives and it is unclear whether all members will recognize this new policy, or how it will be implemented for those countries that do agree to it. The European Commission, which acts on behalf of EU members, has tried to adopt regulations to facilitate the adoption of biotech crops, and its scientific committees have endorsed the safety of many products derived from biotechnology. However, the complexity of the regulatory process of its members has prevented widespread adoption of biotech plants.

In the European Union, regulatory agencies are charged with not approving GE crops or products until it can be stated conclusively that they are safe (Perdikis, 2000). Thus, some proponents of the precautionary principle demand that governments ban the planting of Bt plants until questions about their safety are more fully answered. Already the precautionary principle regulates policy decisions in Germany and Switzerland and "may soon guide the policy of all of Europe" (Appell, 2001). The precautionary principle has been mentioned in the United Nations Biosafety Protocol regulating trade in GM products. Although the situation in the European Union is perhaps the most complex because of its member states, other countries are developing their own processes, some of which favor the use of biotech whereas others do not. The United Kingdom has increased its regulations during the past several years and currently has a moratorium on the commercial release of biotech crops. In Australia, Bt cotton is the only biotech crop grown widely.

In the United States, the Food and Drug Administration (FDA), Environmental Protection Agency (EPA), and Department of Agriculture (USDA) have established regulations that govern the production and consumption of products produced through genetic engineering (NAS, 2000). These agencies work with university scientists and other individuals to develop the data to ensure that these regulations are science based. In the United States, the White House Council on Competitiveness, along with the Office of Science and Technology Policy (OSTP), has articulated a risk-based approach to regulation (Carpenter, 2001). Regulations for GM crops have been developed over time. In the early 1970s, as genetic engineering was being developed, scientists and federal agencies began discussing the relevant safety issues of biotechnology. In 1986, OSTP published its "Coordinated Framework Notice," which declared the USDA as the lead agency for regulation of plants grown for feed, whereas food is regulated by the FDA (NAS, 2000). The EPA regulates pesticides, including microbial pesticides, and in 1992 was given jurisdiction over biotech plants used for pest control, such as corn, cotton, and soybeans. In January 2001, the EPA formalized its existing process for regulating biotech crops and plants that produce their own pesticides or plant-incorporated protectants (PIPs), such as Bt crops. According to the EPA, "If the agency determines that a PIP poses little or no health or environmental risk, they will be exempted from certain regulatory requirements . . . [and] the rules will exempt from tolerance requirements the genetic material (DNA) involved in the production of the pesticidal substance in the plant" (EPA, 2001b). This principle has favored the development and deployment of the current GM plants.

Part of the USDA, the Animal and Plant Health Inspection Service (APHIS) oversees field trials of biotechnology products. Recently, APHIS centralized its biotechnology functions by creating the Biotechnology Regulatory Service (BRS). Over the past 14 years, APHIS has authorized more than 8,700 field trials on 30,000 sites and granted deregulated state to more than 50 transgenic events representing 12 crop species (USDA APHIS, 2002). APHIS has authorized field trials of a number of grasses modified for tolerance to herbicides, salt, drought, and resistance to pests, as well as plants engineered for altered growth or ones used for phytoremediation. Field trials of lumber-producing trees have also been authorized.

As the USDA looks beyond the first generation of biotech crops already in the field, it is examining data on field trials and potential commercialization of ornamentals, trees, turfgrass, fish, and shellfish. As part of the Farm Security and Rural Investment Act of 2002 in the United States, the USDA has been given authority to regulate farm-raised fish and shellfish as livestock under its animal health regulations. As the USDA moves into these new areas, it has requested the

National Research Council (NRC) to undertake a study on confinement of transgenic organisms under experimental and production conditions. The NRC's job is to evaluate the science of techniques for gene confinement and determine what additional data are needed to answer unresolved questions (USDA NASS, 2002).

GROWER ADOPTION OF BIOTECHNOLOGY CROPS

Since the introduction of GM crops in the United States in 1996, virtually no widespread adoption has occurred in the European Union due to moratoriums. Industrialized countries continue to have the majority (76 percent) of the total GM plantings, but Argentina grew nearly 12 million ha of GM soybeans in 2001. In 2000 China and South Africa grew 3.0 million and 0.5 million ha of GM crops respectively (James, 2001b). South Africa has grown GM cotton since 1997 and GM corn since 1998, and both large- and small-scale growers realized higher net incomes due to higher yield and saving on pesticides (Kirsten and Gouse, 2002). In the case of Bt cotton grown by small farmers, growers indicated that the ease of use of the Bt crops was a particular benefit because of the poor infrastructure for pest management (e.g., lack of spray equipment and services), whereas large-scale farmers noted the "peace of mind" and managerial freedom it provided (Kirsten and Gouse, 2002). In March 2002, Bt cotton was cleared for commercialization in India, a major producer of cotton on the world market. In May 2001, the Chinese Ministry of Agriculture and Technology noted that China had developed 47 transgenic plant species (Chapman et al., 2002).

Adoption of GM plants in the United States varies by crop and region (USDA NASS, 2002). In 2002, 71 percent of upland cotton in the United States was biotech (insect or herbicide resistant, or both), but the rate of adoption varied from 52 percent in Texas to 90 percent in Georgia. The national average was 32 percent biotech for corn (insect or herbicide resistant, or both), but the rate of adoption varied from 9 percent in Ohio to 65 percent in South Dakota. Nationally, soybeans averaged 74 percent biotech (herbicide resistant only) and the adoption rates by states (50 to 86 percent) was less variable. The primary reasons given for adopting GM soybeans are ease of weed control and economic benefits.

In a survey in South Dakota, a state that ranks first in the proportion of total cropland devoted to GM corn and soybeans among the major U.S. corn- and soybean-producing states, more than half the growers indicated that per-acre profits increased as a result of using Bt corn, whereas less than half believed their profits increased from using HT corn or soybeans (Van der Sluis and Van Scharrel, 2002). Overall the experience with these GM crops was positive, although half of the growers said their profits were no better or worse. The primary factor determining whether growers would continue to grow GM corn and soybeans was improved pest management. Of those growers concerned about adopting GM crops, concerns included segregating crops, environmental issues, and the potential for receiving a lower price.

ECONOMICS OF USING CURRENT BIOTECHNOLOGY CROPS

Growers may adopt GM plants based on convenience of use or other factors, but their continued use will ultimately be based on economics. HT soybean was the most dominant GM crop in 2000 and constituted nearly 60 percent of the area used to grow GM crops (James, 2000). The adoption rate of HT soybeans was 2 percent in 1996 (James, 2001a) and was expected to reach 74 percent in 2002 (USDA NASS, 2002). The economic benefits in the United States, estimated by Carpenter and Gianessi (2001), were $109 million in 1997, $220 million in 1998, and $216 million in 1999. In Argentina, the only other country growing a substantial area of HT soybean, the estimated economic benefit was $214 million in 1998-1999 and $356 million for 1999-2000 (James, 2001b).

The other dominant HT crop is canola. In 2000 the area used to grow to HT canola was 2.8 million ha, of which about 2.5 million ha was in Canada (James, 2000). Adoption of HT canola in Canada has grown from 4 percent in 1996 to more than 50 percent in 2000 (James, 2001a). The Canola Council of Canada (2001) reports a continuing economic benefit which varied from $5 million when less than 4 percent of the crop was HT to $66 million in 1999 when more than 50 percent was HT. The benefits were largely due to lower (40 percent) weed control costs, lower fuel costs for tractor trips, lower dockage for unclean grain, and better planting times (James, 2000). It should also be pointed out that companies are putting HT into their premium, high-yielding varieties as an incentive for growers to buy HT technology.

Economic analyses using several different methods show a consistent positive economic return to U.S. growers when they use Bt cotton (EPA, 2000). These economic benefits to growers on a national level vary from year to year and from model to model, but range from $16.3 to $161.3 million. Carpenter and Gianessi (2001) stated that Bt cotton farmers in five studies in seven states had a 9 percent yield increase with Bt cotton and that these yield and revenue impacts, if realized over all 4.6 million acres of Bt cotton in 1999, would result in a $99 million increase in revenue. Frisvold and colleagues (2000) provided a more regional-based analysis and estimated that benefits to Bt adopters grew from $57 million in 1996 to $97 million in 1998. Using figures from various sources, James (2000) estimates that the economic advantage of growing Bt cotton in the United States ranges from $80 million to $142 million. In an economic analysis of the distribution of the economic benefits of Bt cotton, Falck-Zepeda and colleagues (2000) calculated that the introduction of Bt cotton created an additional wealth of $240.3 million for 1996. Of this total, the largest share (59 percent) went to U.S. farmers. Monsanto, the developer of the technology, received the next largest share (21 percent), followed by U.S. consumers (9 percent), the rest of the world (6 percent), and the seed companies (5 percent).

In China, the economic benefits of Bt cotton to growers in Liangshan county of Shandong province were $930/ha in 1998, and the estimated average benefits were about $250/ha in 1998-2000 (Jia et al., 2001). In a larger survey of 283 cotton farmers in Northern China in 1999, Pray et al. (2001) reported that the cost of production for small farmers was reduced by 20 to 33 percent by using Bt cotton depending on variety and location, and "the net income and returns to labor of all the Bt varieties are superior to the non-Bt varieties." Pray et al. (2001) estimated that the national benefit of Bt cotton in China is $139 million annually. Due to the technology fee for the seed, Australian growers are saving little on costs, but they have adopted Bt cotton for the improved certainty of yields and to reduce concerns about environmental contamination with insecticides (Forrester, 1997). The economic benefits in China and Australia may be lower because *Helicoverpa* species, which are the main pests in those countries, are at least tenfold less sensitive to Cry1A than *H. virescens*, the key pest in the United States.

The economic advantage of Bt corn is not as dramatic as that of Bt cotton. Because most growers do not presently use insecticides to control European corn borer (ECB), but accept the losses it causes, it is more difficult to assess the impact of Bt corn. However, growers' buying habits in the United States appear to confirm some economic advantage in growing Bt corn because the percentage of Bt corn in the total crop has grown from less than 1 percent in 1996 to 26 percent in 2001 (James, 2001b). Although ECB causes significant yield loss, infestation levels and resulting losses are inconsistent from year to year, and, therefore, it is difficult to predict whether control is needed. With Bt technology, growers must plant the crop and incur pest-control costs before knowing whether they need it. On the other hand, if growers do not use the Bt crop, they will not be able to get the same level of control. Thus, they must perform a risk-benefit analysis early in the season. Carpenter and Gianessi (2001) estimate an average net benefit to growers of $18 per acre in 1997 (a year of high infestation) to a loss of $1.81 per acre in 1998 (a year of low infestation and low corn prices). Using another model, EPA estimates indicate a net benefit of $3.31 per acre on 19.7 million acres of Bt corn planted in 1999, or a national benefit of $65.4 million (EPA, 2000). Carpenter and Gianessi (2001) estimate that in "10 of the 13 years between 1986 and 1998, ECB infestations were such that corn growers would have realized a gain from planting Bt corn."

Virus-resistant transgenic crops constitute only a small percentage of the overall GM market at present, with papaya and squash as the only two crops on the market. In the late 1980s and early 1990s, the $45-million Hawaiian papaya industry was devastated by the papaya ringspot virus (PRSV) and production fell from 58 million pounds in 1992 to 24 million in 1998. In 1998 two PRSV-resistant varieties were commercialized and released to Hawaiian farmers and, for the first time since 1992, production increased. However, because the primary market for Hawaiian papaya was Japan, where GM papaya has not been approved, it is being sold primarily in the United States.

The development and commercialization of agricultural biotechnology has enhanced productivity (Shoemaker et al., 2001), but has economic consequences in the form of commodity prices, supply and demand, and international trade. Using corn and soybean production in the United States as an example, Barkley (2002) constructed an economic analysis to calculate the impact of GM crops on market prices, production, domestic demand, and exports of corn and soybeans produced in the United States. Barkley's results document that producer adoption of GE crops results in an increase in supply of corn and soybeans, price reductions, and increases in domestic supply and exports. Current adopton rates in the United States are 32 percent for corn and 74 percent for soybeans. Even if worldwide adoption of these two crops were to take place, prices of these two commodities would be reduced by less than 2 percent. For producers of these crops who are using biotechnology to reduce production costs, the small decrease in global grain prices is likely to be more than offset by cost-reducing gains in efficiency. Thus, adoption rates in the United States are likely to continue to increase, leading to slightly lower global prices and relatively larger gains in exports (Barkley, 2002).

THE CONCEPT OF SUBSTANTIAL EQUIVALENCE AND ITS IMPACT ON PRODUCT LIABILITY AND INTERNATIONAL TRADE

Countries such as the United States and Canada focus on determining the safety of the product, rather than the process of how that product was developed. This is usually referred to as *substantial equivalence*. The Food and Agriculture Organization (FAO) of the United Nations and the World Health Organization (WHO, 2000) committee recommends:

> GM foods should be treated by analogy with their non-GM antecedents, and evaluated primarily by comparing their compositional data with those from their natural antecedents, so that they could be presumed to be similarly acceptable. Only if there were glaring and important compositional differences might it be appropriate to require further tests, to be decided on a case-by-case basis.

In contrast, the European Union established a separate regulatory framework that requires a premarket approval system for GM foods, as well as regulations for how GM crops are grown. Thus, the main difference between the European Union and United States/Canada is that the former focuses on the process whereas the later focuses on product (Hobbs et al., 2002).

In the United States, food labels reflect composition and safety, not the way the food is produced. At present, foods derived through biotechnology do not require labeling because they have been judged to have the same nutritional content and no changes in allergens or other harmful substances. In addition, some products, such as oils derived from biotech crops, are indistinguishable from those derived from nonbiotech crops. If the presently available biotech foods were to require labels, it would not be on the basis of nutrition or food safety (the current requirements) but on the way they were produced. Conventionally produced agricultural products do not require labeling describing how they were produced. It is estimated that if foods were certified to be biotech free, it would increase the cost of the food because the product would have to be followed (traced) from the field to the market. The situation is far more complex if processed foods are to be certified. A processed food (e.g., pizza) may contain dozens of ingredients and to certify it as biotech free would require certification of each ingredient. It is unclear how biotech products would be segregated in a complex food system and who would pay for the additional costs. However, future biotech products are expected to have improved nutritional value, and in the United States they will have to be labeled to that effect. A fundamental question is whether labeling would help consumers make an informed choice about the safety or nutritional value of their foods. The U.S. regulations on labeling stand in stark contrast to those of some other countries. The European Union has adopted a policy of mandatory labeling for all foodstuffs containing GM products above a 1 percent threshold. Japan has a similar policy, except at a 5 percent threshold.

In July 2001, the European Commission unveiled its proposals on labeling and traceability of foods containing GM products. The rules require the labeling of all foods and animal feed derived from GM crops and, in the case of processed goods, that records be kept throughout the production process, allowing GMOs to be traced back to the farm (Chapman et al., 2002). Reiss (2002), the former ethicist on the British government's Advisory Committee on Novel Foods and Processes, expressed concern on the proposals from the European Commission on traceability and labeling of GM crops and the food and feed products derived from them. His main concern is that labels will be required regardless of whether detectable levels of DNA or proteins are in the product, so it really is the process that is being regulated. Reiss (2002) points out that much of the purported justification for labeling is based on faulty surveys of consumers, in which they are not asked about the potential consequences of labeling, such as practicality and cost to the consumer.

Gaisford and Chui-Ha (2000) suggest that objections to GM food products can come from environmental and ethical concerns, as well as questions about food safety. Thus, the challenge for policymakers is to design regulations for various audiences who have different values on a variety of aspects relating to food (Hobbs et al., 2002). This becomes even more complex because of cultural differences in different countries, as well as the regulators for international trade. Hobbs and colleagues (2002) provide an excellent overview of the trade regulations affecting GM foods. The World Trade Organization (WTO) was formed as a means to deal with multilateral trade negotiations through regulations. The WTO administers a number of voluntary agreements between nations, including the Sanitary and Phytosanitary Measures (SPS) that require "the best scientific information" to restrict trade between any members subscribing to WTO agreements. Another agreement, the Technical Barriers to Trade (TBT), deals with packaging, labeling, and product standards, and is not subject to scientific principles. According to Hobbs and colleagues (2002), the TBT allows importers to require labeling if a product, not the process by which it is produced, is novel. Therein lies the rub between the European Union and the United States and Canada, because it comes down to the perceived safety of GM foods and whether SPS or TBT applies. If a product falls under SPS, then a scientific risk analysis needs to be conducted. If the product is considered safe, then the TBT applies. At this time is unclear whether GM labeling is an SPS or a TBT issue, and this contributes to international conflict.

To make the situation even more complex, the WTO employs the most-favored-nation principle, which requires countries to treat similar products from all countries in the same manner. The question may arise whether a GM and non-GM product are the same, especially when the GM product contains no detectable foreign protein or DNA, as would be the case with oils derived from GM plants. According to Hobbs and colleagues (2002), the debate on GM products has led to a shift in trade protection because regulations are being developed to restrict trade based on consumer and environmental interests for products that are perceived to be harmful if imported, instead of protecting producers in one country from products created in another. Thus, the SPS and TBT agreements, which were set up to protect producers, are now being used by groups opposed to biotechnology. Although consumer preference should be of some importance in trade regulations, the current WTO agreements are not structured with this in mind. In July 2003, the *Codex Alimentarius* Commission (FAO 2003) met with 125 countries and adopted an agreement on how to assess the risks to consumers from foods derived from biotechnology, including genetically modified foods. These guidelines lay out broad general principles intended to make the analysis and management of risks related to foods derived from biotechnology uniform across Codex's 169 member countries. Provisions of the guidelines include premarket safety evaluations, including allergenicity, product tracing for recall purposes, and postmarket monitoring. The guidelines cover the scientific assessment of DNA-modified plants, such as maize, soy, or potatoes, and foods and beverages derived from DNA-modified microorganisms, including cheese, yogurt, and beer. The guidelines are considered to be based on sound science and have the support of the biotechnology companies. Meanwhile, clashes continue over the production and consumption of GM foods in the various marketplaces of the world. The regulatory regime governing international trade has been slow to adapt its rules to the movement of biotech products—a result of the slow pace of WTO negotiations (Hobbs et al., 2002). Therefore, nations have been largely developing their own regulatory framework, some of which conflict with existing WTO commitments.

A prime reason for the clashes is the difference in consumer attitudes in various parts of the world. In general, Americans are more confident about the safety of their food supply and trust their government regulations, which has led them to be more accepting of GM foods (Wolf et al., 2002). A recent survey (Fresh Trends, 2001) noted that American consumers accepted modified food to be more resistant to plant diseases and less reliant on pesticides (70 percent), help prevent human diseases (64 percent), improve nutrition value (58 percent), improve flavor (49 percent), and extend shelf life (48 percent). By contrast, EU consumers generally view GM foods negatively, perhaps due to the politically active Green Movement or recent incidents of mad cow disease (Pennings et al., 2002) and other safety concerns that have plagued Europe. Wolf and colleagues (2002) found that half of the consumers in the United States were familiar with GM food, compared to only 28 percent of the consumers in Italy, and that consumers in the United States were much more likely to consume GM food than their European counterparts. Respondents also had different opinions on the appropriate sources of information for GM foods, again reflecting cultural differences. In order to be effective, communication efforts should be based on insights in the formation and impact of consumer beliefs and attitudes regarding GM foods and targeted to specific consumer segments. How this is accomplished in a pluralistic society such as the European Union (or even the United States) will be a particular challenge, and even more so in developing countries, such as Colombia (Pachico and Wolf, 2002), where education levels are lower.

A recent survey in the United States (IFIC, 2002) found that American consumer support for biotechnology is holding steady or even increasing. This survey, which asked questions in a risk-and-benefit comparison, found that 71 percent of the respondents said they would likely buy GM produce that was protected from insect damage and thereby required fewer insecticide applications. Overall awareness of biotechnology also remains high in the United States (72 percent), and the majority (59 percent) of Americans support the FDA's policy that the GM products available at present should not be labeled because they are substantially equivalent to their non-GM counterparts. This survey clearly points out the difference in public policy attitude between the United States and the European Union, i.e., regulation of the process (European Union) versus the product (United States).

In the case of GM products, opponents of labeling are concerned that identification of a product as GM will have negative connotations about the wholesomeness of the product. The USDA's Economic Research Service notes that costs for labeling would be significant because it would require identity preservation. The higher costs would affect all consumers and thereby be similar to a regressive tax, because the poor spend a proportionately larger share of their income on food than do high-income households (Huffmann et al., 2002). Although the exact costs of labeling are not known, some studies have estimated that it could be as high as 15 percent. At least one study (Huffmann et al., 2002) indicates that a voluntary, rather than mandatory, labeling policy would be more effective in the United States.

The situation for the U.S. export market is more complex, however, since some markets, such as the European Union and Japan, respectively, have a 1 percent and 5 percent tolerance for biotech-commingled products. The identify preservation (IP) system currently employed in the United States can handle the 5 percent tolerance for Japan, and the costs for IP are passed along to Japanese consumers. However, the current U.S. system is not designed to handle the EU's 1 percent tolerance, and it is questionable whether the costs for more stringent IP requirements in the EU will be borne by EU consumers (Lin, 2002).

A diversity of opinions on labeling exists even in the EU. Meanwhile, editorials in major U.S. newspapers regularly discuss the European Union's refusal to license new GM crops "even though Europe's own health commissioner says the ban violates international trade rules" (*The Washington Post*, 2002). According to this editorial, Europe is out to "protect their own producers against biotech-powered Americans," and labeling is not invoked in Europe when its cheeses made with GM enzymes are noted. Faced with this dilemma the editorial recommends bringing suit against the European Union at the WTO.

FOOD SAFETY ISSUES
OF BIOTECHNOLOGY PRODUCTS

It is impossible to provide consumers assurance of absolute zero risk for food products, largely owing to the inadequacy of methods to screen for novel and previously unreported toxicity or allergenicity. However, the zero-risk standard that is applied to this new technology far exceeds the standard used for novel crops produced by conventional methods. In 1992 the U.S. FDA provided a decision tree for assessing the safety of biotechnology-derived food products based on a risk analysis related to the characteristics of the products, not the process by which they were created. In practice, a series of specific questions are addressed:

- Is the DNA safe to consume?
- If an antibiotic marker was used to create the product, is it safe?
- Are the newly introduced proteins safe to consume?
- Have the composition and nutrition value changed?

- Are there changes in important substances?
- In what forms will the food products be consumed?
- Do the newly introduced substances survive processing and preparation?
- What is the expected human dietary exposure? (Chassy, 2002)

Food safety assessment depends heavily on assumptions contained in the concept of substantial equivalency, but some of these assumptions can be examined through the scientific process for the GM plants presently available, as well as those that may be utilized in the future.

In the case of Bt plants, several types of data are required to provide a reasonable certainty that no harm will result from the aggregate exposure of these proteins (see review of data in Shelton et al., 2002). The information is intended to show that the Bt protein behaves as would be expected of a dietary protein, is not structurally related to any known food allergen or protein toxin, and does not display any oral toxicity when administered at high doses. U.S. regulatory agencies do not conduct long-term studies because they believe that the instability of the protein in digestive fluids eliminates this need. The in vitro digestion assays that are used attempt to confirm that the Bt protein is degraded into small peptides or amino acids in solutions that mimic digestive fluids, but the assays are not intended to provide information on the toxicity of the protein itself. Acute oral toxicity is assessed through feeding studies with mice using a pure preparation of the plant-pesticide protein at doses of >5,000 mg/kg body weight. None of the Bt proteins registered as plant pesticides in the United States have shown any significant effect (Betz et al., 2000; EPA, 2000; WHO, 2000). The potential allergenicity of a Bt protein can be examined through in vitro digestion assays, but further assessment is done by examining amino acid homology against a database of known allergens. None of the presently registered Bt proteins have been demonstrated to be toxic to humans, nor have they been implicated to be allergens. Furthermore, they do not contain sequences resembling relevant allergen epitopes (Chassy, 2002).

Because of the StarLink situation (registration of this Cry9C product for animal feed only, although it became commingled with products for human consumption), the EPA (1999, 2001b) addressed more fully the allergenicity concerns with Cry9C. Their report stated that "while Cry9C has a 'medium likelihood' to be an allergen, the combination of the expression level of the protein and the amount of corn found to be commingled poses a 'low probability' of sensitizing individuals to Cry9C" (EPA, 1999). Later studies by the Centers for Disease Control (CDC) indicated that none of the people claiming they had developed allergenic reactions to Cry9C had actually done so (CDC, 2001).

Heat studies have also been conducted because many of these Bt plant products are processed into foods. A key issue is that not all foodstuffs prepared from GM crops are GM. Protein and DNA are destroyed during the processing of highly refined foodstuffs, such as oils and sugars. This is especially true for cottonseed oil, which must be heavily refined to remove toxic secondary plant compounds. Not only does cottonseed oil contain no DNA or protein, no consistent difference exists between GM and non-GM cottonseed oil in compositional analyses (Food Standards Australia New Zealand, 2002). Cry1Ab and Cry1Ac became inactive in processed corn and cottonseed meal, but Cry9C was stable when exposed to simulated gastric digestion and to temperatures of 90°C (EPA, 1999), and was therefore not permitted for human consumption, although it was allowed for animal consumption, a decision that led to the StarLink situation.

In contrast to concerns about toxicity and allergens from GM, clear evidence exists for health benefits from Bt corn (Shelton et al., 2002). *Fusarium* ear rot is a common ear rot disease in the Corn Belt, and the primary importance of this disease is its association with mycotoxins, particularly the fumonisins, a group of mycotoxins that can be fatal to horses and pigs and are probable human carcinogens (Munkvold and Desjardins, 1997). Field studies have demonstrated that Bt hybrids experience significantly lower incidence and severity of *Fusarium* ear rot because of superior insect control, and yield corn with lower fumonisin concentrations than their non-Bt counterparts (Munkvold et al. 1999).

Because the majority of corn worldwide is fed to livestock, questions arise about their suitability as animal feeds. In a study using Bt corn silage on the performance of dairy cows, the authors found no significant differences between Bt and non-Bt corn hybrids in lactational performance or ruminal fermentation (Folmer et al., 2000). A summary of studies on Bt crops fed to chicken broilers, chicken layers, catfish, swine, sheep, lactating dairy cattle, and beef cattle was compiled by the Federation of Animal Science Societies (2001). In a review using these studies, Faust (2000) concludes that there are "no detrimental effects for growth, performance, observed health, composition of meal, milk, and eggs, etc."

For HT crops, the EPSPS enzyme is thermally labile, rapidly digestible in a gastric environment, not glycosylated, and possesses no amino acid sequences homologous to known food allergens (Carpenter et al., 2002). The European Commission's Health and Consumer Protection Directorate-General (2002) reviewed an HT maize with the modified EPSPS and noted that the line was as safe as grain from conventional lines. In a review of the safety issues associated with DNA in animal feed derived from GM crops, Beever and Kemp (2000) examined the range of issues from protein safety to the uptake and integration of foreign DNA to the potential for antibiotic resistance-marker DNA, and concluded that "consumption of milk, meat and eggs produced from animals fed GM crops should be considered as safe as traditional practices." Reports issued by the Institute of Food Technologists (2000) have similar conclusions and state that "biotechnology processes tend to reduce risks because they are more precise and predictable than conventional techniques."

In addition to EPSPS and the proteins from Bt, concerns have been raised regarding the use of antibiotic resistance marker genes used in the development of some GM plants. Because antibiotic resistance strains have medical importance, studies have been undertaken to evaluate the risk in humans. In an analysis of the published reports, Smalla and colleagues (2000) suggest that antibiotic resistance is already common and would not be affected by GM crops carrying these markers. Regardless of the science, but more for public relations, the biotechnology industry is looking for other breeding markers.

Questions arise on whether genetic engineering poses risks that cannot be anticipated. For example, once a gene is inserted into a plant, will it be stable and not interact with other genes in a fashion that causes unwanted changes in the plant? Examples of this could include silencing of other genes, functional instability, and a range of other pleiotropic effects (e.g., Benbrook, 2000—but also see the reply to these issues by Beachy et al., 2002). Although these possibilities should be of concern, several international scientific organizations have concluded that the risks of food derived from biotechnology are no different from those associated with new varieties of plants produced through conventional plant breeding (see review by Chassy,

2002; Society of Toxicology, 2002). Reports issued by the Institute of Food Technologists (2000) go further and state that "biotechnology processes tend to reduce risks because they are more precise and predictable than conventional techniques." Still, the debate continues and provides interesting reading, and sometimes sharp dialogues, between opponents (Beachy et al., 2002)

FRAMEWORK FOR EVALUATING ENVIRONMENTAL EFFECTS OF CURRENT BIOTECHNOLOGY CROPS

Ecological risk assessment faces the challenge of defining adverse ecological effects in terms of testable hypotheses within realistic temporal and spatial constraints (Nickson and McKee, 2002). Environmental risk assessment should be a tool for making objective decisions by assessing what is known, learning what is needed, and weighing the costs of mitigating risk or reducing uncertainty against the benefits. In this context it is also important to evaluate the risk with the new product to that of some reference practice, so as to make a comparative analysis in accordance with the precautionary principle as articulated by Goklany (2002). One of the problems encountered with environmental risk assessment is what starting point should be used? Because agriculture has evolved over thousands of years and the pace of its deployment is now so rapid (especially for GM crops), there is no baseline of knowledge from which to monitor potential adverse impacts (Benbrook 2000). Despite these difficulties, scientists and policymakers need to try and decrease our uncertainty about the environmental impact of any technology, including those that have been used for decades or centuries.

Nickson and McKee (2002) suggest four steps in the process for environmental risk assessment. First, a need exists to clearly formulate the problem, develop a conceptual model, and develop an analysis plan. This should be followed by an evaluation of the exposure and the ecological effects or responses through systematic data collection. Third is the need to develop a risk characterization based on the magnitude of the hazard and the probability of exposure. The fourth step is to develop a risk-management plan to reduce, mitigate, or manage determined environmental risks. This model also suggests the need to acquire more data and monitor the results of the process on an ongoing basis. As additional traits are introduced into plants and other organisms, they should follow the same model approach, but will require data particular to the introduced trait and its interaction with the environment.

Within this framework, judgments about the potential risk of a GM product and research needed to address the question of risk should be made using a tiered approach based on the novelty of the new GM organism and how it may interact within the environment. Within this context, the currently registered GM pest-management plants should be evaluated in several areas, including gene flow, shifts in population structure, effects on beneficial and other nontarget organisms, development of resistance, and pesticide-use patterns.

Prior to their introduction, the present GM plants were evaluated by the EPA on a number of environmental issues. Data from industry and university scientists were used by the EPA for its decisions, but it is fair to say that many of these studies were conducted on a small scale and over a relatively short period of time. Such studies provide some ecologically relevant information, but have limitations because they are not the long-term studies suggested by some. Such efforts are limited by funding as well as the interest in moving products into the market on a commercial scale in order to determine their effects more fully. Although some may consider this risky and a violation of the precautionary principle, others would argue that the initial evaluations, performed in greenhouses and small-scale trials required by APHIS before commercialization, allowed a body of data to be evaluated prior to commercialization. They would further argue that risk assessment is an ongoing process and that commercialization is needed to assess long-term effects, such as changes in pesticide-use patterns.

Risk assessment is performed to reduce the uncertainty of a particular practice, but the critical questions are how much uncertainty is acceptable before something is implemented, what is the financial and human cost needed to obtain an acceptable level of certainty, and what is the risk of not implementing a new practice but continuing with the old? In the United States, the USDA is reported to spend little on risk assessment to biotechnology (Snow and Palma, 1997), but this has doubled recently. In the European Union, funding for ecological risk assessment has increased dramatically in recent years due to the introduction of GM crops. Long-term ecological studies with high degrees of statistical power are needed to answer many of the questions posed. As noted by Marvier (2001), the lack of power in ecological studies may lead to the misinterpretation of a hazard. A panel of biologists and agricultural scientists convened by the U.S. National Academy of Sciences examined the federal government's policies on approving crops produced through biotechnology. The report (NAS, 2002) noted that the "standards being set for transgenic crops are much higher than for their conventional counterparts." Furthermore, it stated that although the USDA had not approved any biotech crops that have posed undue risk to the environment, they did recommend changes in the review process to ensure the environmental safety of products that could be introduced in the future. These recommendations included increased public comment and broader scientific input to enhance the regulatory process.

GENE FLOW IN TRANSGENIC CROPS

The GM plants currently enjoying wide usage (corn, soybean, and cotton) do not employ methods of controlling the transgene in the subsequent crop, therefore the transgene(s) may theoretically flow and be expressed in subsequent generations, provided specific conditions are met. This potential for gene flow from transgenic crops can have different ecological, economic, and political ramifications for the long-term use of GM plants.

Nearly all of the main important crops grown worldwide are capable of hybridizing with wild relatives (Ellstrand et al., 1999) and have done so for centuries. Snow (2002) suggests that GM raises additional concerns because it not only enables the introduction into ecosystems of genes that confer novel fineness-related traits but also allows novel genes to be introduced into many diverse types of crops, each with its own specific potential to outcross. Thus, newly introduced genes could bring along new phenotypic traits, such as resistance to insects, diseases, herbicides, or specific growing conditions. Required research into the factors important to understanding the consequences is often lacking. The movement of transgenes from biotech crops represents challenges to the continued use of GM crops, and public visibility and scientific interest in this area has increased. More important than the direct movement of such transgenes are their potential consequences when outside the biotech crop. According to Wilkinson (2002), four categories will help in creating a structured risk analysis leading to the quantification of the risks posed by a GM cultivar. The first is generic information that is likely to hold true for all GM lines of a particular crop, such as whether it is self-pollinating. A second category relates to generic information on the class of transgenes, such as those that enable a plant to be tolerant to a herbicide. Third, it is important to understand the specifics of a certain transgene or construct, such as whether it is produced only at certain times in the crop. Fourth, information on the specific transgene-plant genotype is needed. The initial goal of current research on risk assessment focuses on the first three categories, and their potential to lead to environmental change.

The processes by which transgenes can move from a GM crop and lead to environmental changes are complex and numerous, and will differ depending on the crop-transgene locality and environment. With the GM crops available presently, two mechanisms allow transgenes to move between plant species: dispersal in viable pollen or dissemination in seed. Gene flow through pollen depends on the distance between the donor and recipient plant, the amount of pollen produced and its longevity, the method of dispersal (wind, insects, or other animals), and climatic conditions. Gene flow through seed can occur with seed left in the field, dispersal during transportation, by human intention, or by animals. Predicting all possible outcomes of transgene movement and environmental effect is impossible, and a structured approach should be used with each stage in the process being examined in a progressive manner. Given all the possible factors involved in gene flow, stabilization of the transgene in the non-GM plant, and fitness advantages/disadvantages, Wilkinson (2002) suggests a structured approach in which each stage of a crop recipient is evaluated. As a first step, he suggests one should examine the potential for a hybrid to be formed. In theory the hybrid could be formed by a transgene moving to a non-GM crop, to a feral crop, to a conspecific wild or weedy relative, or to a weedy relative of another species. However, the risk approach should also include information on the likelihood of this occurring under natural conditions and the ability to detect such hybrids using statistically rigorous methods. But hybridization is little more than a step and should be followed by examination of whether the transgene becomes fixed in the population, its potential for a fitness advantage, and its potential for spread in the population. The importance of the transgene will depend on the phenotype conferred, but it is impossible to foresee the effect without a priori knowledge of the function of the gene and detailed knowledge of the ecology of the recipient feral plant (Wilkinson, 2002).

For most questions about the ecological and agronomic consequences of gene flow, our ability to quantify and predict outcomes is still rudimentary, and it is much easier to rule out unwanted scenarios for certain low-risk crops (Snow, 2002). For example, over the cultivated range of many crops there are no cross-compatible wild relatives (Wilkinson, 2002). Thus, the risk of gene flow in a crop's center of origin (e.g., maize in Mexico) will be different than in areas where no such relatives exist (e.g., United States, Canada, and Europe). Considerable publicity has surrounded the Quist and Chapela article (2001), which claimed that transgenic DNA sequences were present and had introgressed into native Mexican corn. Although the editor of *Nature* (2002) eventually disavowed the findings, it would be surprising if, over time, the transgenes were not found to have spread. The question is whether they would have any deleterious consequences. Meanwhile, this event highlighted the difficulty of regulating GM crops and provided antibiotechnology forces with new political ammunition against GM crops. In the case of those plants registered in the United States (corn, cotton, and potatoes), it is unlikely that they will pass their traits to wild relatives because of differences in chromosome number, phenology, and habitat (EPA, 2000). The only exception is cotton in Florida and Hawaii, where feral populations exist of related *Gossypium* species. The EPA has prohibited or restricted the use of cotton in these areas.

GENE CONTAINMENT STRATEGIES FOR BIOTECHNOLOGY CROPS

Cultural practices, deployed in time or space, can reduce the flow of pollen and hence gene flow. For example, GM crops can be bordered by non-GM crops that will capture pollen as it moves away from the GM crop, or borders may serve to isolate the crop from pollen moving into the field. Barriers and other isolation tactics are already being used as standard practices for seed production. However, the stricter requirements for GM crops may necessitate even larger barriers in time or space. These cultural practices need to be tailored to each crop-transgene system. In the future, it is likely that regulations will be developed that require the use of molecular gene containment systems to avoid adverse environmental effects, but at present none work across all crops. However, molecular technologies offer the opportunity to alter gene flow by interfering with flower pollination, fertilization, and/or fruit development (Daniell, 2002). Male sterility is already used in breeding canola to prevent outcrossing from GM canola to weeds, but seeds producing male sterile GM crops by cross-pollination from weeds may be problematic because seeds of the hybrids will produce fertile pollen carrying the GM trait (Daniell, 2002). Maternal inheritance of cytoplasmic organelles through chloroplast is being exploited as a means to prevent the transmission of transgenes through pollen, and this technique has been demonstrated in several plant species. Producing plant seeds that are sterile was a promising technique that met its demise when it was dubbed "Terminator Technology" by opponents of biotechnology. Although most farmers in wealthy countries do not save seed to replant the next year and could have benefited from such technology, saving seed is a common practice in developing countries. Other molecular techniques are being developed, but full utilization of any of them will depend on a more complete understanding of the regulatory and structural genes involved in pollen and embryo functions.

ECONOMIC CONSEQUENCES OF GENE FLOW FROM BIOTECHNOLOGY CROPS

Smyth and colleagues (2002) note several examples that gene flow from GM plants created economic liability by affecting the GM tolerance level of a crop. In 1999 the Swiss Department of Agriculture found non-GM corn to be "contaminated" with foreign genes, and crops were destroyed and compensation was paid to growers. In 2000 the European Union found that a shipment of canola seed contained 0.4 percent unapproved GM traits as a result of gene flow from foundation GM canola produced in Canada. Although the contamination was less than 1 percent, a figure often cited as a tolerance level, France ordered 600 ha to be plowed under. Another example occurred in 1999, when the European Union detected the presence of a GM protein not yet approved in the European Union in a shipment of honey from Canada and rejected the shipment. The European Union later went on to ban all shipments because of the inability of Canadian producers to guarantee that such non-EU proteins would not be present in future samples. Although trade rejections may be the result of other factors such as protectionism, GM contamination will continue to be an issue between countries that have different regulations and standards on GM crops. Within a country, gene flow may also affect markets. Smyth and colleagues (2002) note that the "introduction of transgenic herbicide-tolerant canola in western Canada destroyed the growing, albeit limited, market for organic canola."

An overview of recent litigation with GM crops is contained in an article by Jones (2002). Several suits have been brought against biotech companies by farmers, but perhaps the most famous was the discovery of StarLink corn in the human food supply. The protein Cry9C contained in StarLink was registered for animal consumption only but became commingled with other corn and resulted in at least nine class action suits against Aventis CropScience, the producer of StarLink. Some of these claims are for lost revenue due to reduced corn prices because of the incident. In Canada, suits have been brought against Monsanto and Aventis for causing the loss of organic canola and wheat markets. One of the more intriguing lawsuits is that brought by Monsanto against Percy Schmeiser in Canada for growing Roundup Ready canola. In this case, Schmeiser did not have a license from Monsanto to grow its canola but claimed that the plant appeared on his farm by accident. Schmeiser lost the case but plans to appeal (Jones, 2002).

EFFECTS OF BIOTECHNOLOGY CROPS ON NONTARGET ORGANISMS

Bt Crops

Because Bt has a narrow spectrum of activity against insects and is nontoxic to humans, substitution of Bt crops for broad-spectrum insecticides would likely result in conservation of natural enemies and nontarget organisms, decreased potential of soil and water contamination, and benefits to farm workers and others likely to come into contact with these insecticides. As the agency charged with regulating the use of Bt plants in the United States, the EPA has served as the clearinghouse for studies examining the effects on non-target organisms. In addition to the EPA's reports, additional synthesis studies have been published by Snow and Palma (1997), Traynor and Westwood (1999), and Wolfenbarger and Phifer (2001).

Prior to the registration of the first Bt crop in 1995, the EPA evaluated studies of potential effects of Bt endotoxins on a series of nontarget organisms, including birds, fish, honey bees, ladybugs, parasitic wasps, lacewings, springtails, aquatic invertebrates, and earthworms (EPA, 1995). Organisms were chosen as indicators of potential adverse effects when these crops are used in the field. These studies consisted of published reports as well as company reports (Shelton et al., 2002). Their focus was primarily on toxicity to the species tested because, unlike some synthetic insecticides that bioaccumulate, no data suggest that Bt proteins concentrate in the species and harm them. From its review of existing data, the EPA concluded that "no unreasonable adverse effects to humans, nontarget organisms, or to the environment" existed (EPA, 1995). Since the commercialized use of Bt plants in the field, however, some highly publicized reports have suggested negative impacts on nontarget organisms. Several of these reports are worth noting. A small preliminary lab study indicated potential risk of monarch butterflies to Bt pollen (Losey et al., 1999). This report received tremendous news coverage but was also severely criticized in scientific circles for its methods and interpretation (Shelton and Sears, 2001). More detailed laboratory studies and a series of field studies have shown that the risk to monarch butterfly populations in the field is "negligible" (Sears et al., 2001).

Reports indicating that consumption of corn pollen affects the development and mortality of lacewing larvae have created discussion focusing on the compatibility of Bt plants and biological control (Shelton et al., 2002). Hilbeck and colleagues (1998) reported increased mortality and prolonged development when lacewing larvae were reared on two pests of corn. Their experimental design did not permit a distinction between a direct effect due to the Bt protein on the predator versus an indirect effect of consuming a suboptimal diet consisting of sick or dying prey that had succumbed to the Bt protein. The authors also noted that their study was unrealistic because the pests "will almost completely be eradicated" by the Bt plants. Although no relevent conclusions can be drawn from this and other similar studies, they do show the difficulty in conducting laboratory studies on tritrophic interactions that have relevance in the field. Interactions in the laboratory, although dramatic, may not be realistic in the field. Likewise, testing only a single factor in the laboratory may not produce the subtle effects that may arise in the field.

A third set of reports suggested that Bt exudates from the roots of 13 Bt corn hybrids could accumulate in the soil during plant growth as well as in crop residues (Saxena et al., 1999). To assess the effects of Cry1Ab toxin released in the root and from biomass on soil organisms, researchers introduced earthworms into soil grown to Bt and non-Bt corn or amended with biomass of Bt or non-Bt corn. Although the protein was present in the casts and guts of worms in the Bt treatments, there were no significant differences in mortality or weight of the earthworms, nor in the "total numbers of nematodes and culturable protozoa, bacteria (including actinomycetes), and fungi between rhizosphere soil of Bt and non-Bt corn or between soils amended with Bt or non-Bt biomass" (Saxena and Stotzsky, 2001).

Considerable research has been conducted to assess changes in composition of natural enemy populations in Bt and non-Bt crops. Several studies have shown lower natural enemy populations in Bt crops, although this is attributed to reduced populations of the hosts they would have fed on due to the effectiveness of the Bt crops (Carpenter et al., 2002). It is generally recognized that natural enemy populations are more negatively impacted by use of alternative insecticide treatements. Although the data to date do not indicate striking problems with Bt proteins on nontarget organisms, they point out the difficulty in working in a complex environment. Studies often focus on single organisms under specific environmental conditions and over an often short period of time. Under these conditions, the power to test for differences is relatively low, and longer-term and more complex studies are needed to ensure the integrity of important nontarget organisms.

Herbicide Tolerant (HT) Crops

HT crops allow the use of minimum or nontill farming, which, in turn, will have an effect on the microbial community in the soil, and this effect is generally considered positive. Stabilizing the soil structure and the soil microbial community through minimal and no-till practices will enhance nutrient recycling, reduce soil erosion and compaction, and lessen water runoff (Carpenter et al., 2002). Apart from this indirect effect, the use of glyphosate has increased dramatically with the introduction of HT plants, and studies have been conducted to determine the consequences. A report by Haney and colleagues (2000) suggested that glyphosate did not affect the soil microbial biomass; however, other reports suggest that glyphosate may affect symbiotic bacteria, which provide soybeans with much of their nitrogen requirement, and thus affect soil fertility (Carpenter et al., 2002). Other reports indicate that use of HT soybeans did not have detrimental effects on beneficial insect populations (Janinski et al., 2001), although it is assumed that good weed control would decrease the overall biodiversity within the field regardless of whether the weed control came from use of HT crops or mechanical cultivation. Additional studies (reported by Carpenter et al., 2002) indicate no adverse affects on earthworms or birds. However, again showing the complexity of agricultural systems, a study reported in the United Kingdom (Watkinson et al., 2000) examined the effect of no-till farming, a practice accomplished easily with HT crops. Because of the decrease in weeds and their seeds when using HT technology, a mathematical model indicated a decline in a particular species of bird that feeds on weed seeds, but again this would occur with any weed-management tactic.

When one herbicide is used extensively, the population structure of the weed community may change. As reported by Carpenter and colleagues (2002), two weed species not very susceptible to glyphosate are becoming increasingly problematic in Iowa, and a similar situation has occurred in Kansas. The authors suggest a possible solution to this problem is the use of crop rotation with biotechnology-derived crops tolerant to different herbicides or use of conventional weed-management tactics, or even stacked HT traits.

Virus-Resistant Crops

GM virus-resistant crops presently are grown on very limited areas and the molecular mechanisms by which they display resistance to viruses can be complex. The potential risks of GM virus-resistant plants are relatively small, and once the potential risks have been identified and their underlying mechanisms understood, in many cases the source of risk can be eliminated (Tepfer, 2002). On a molecular basis, this means additional research is needed on virus complementation, heterologous encapsidation, and synergy in the particular virus/plant system. In addition to the virus/plant system, another level of complexity can exist because some of these viruses are insect transmitted. Changes in the molecular nature of the virus may change its ability to be transmitted by its insect vector. As in insect-protected and HT plants, however, perhaps more emphasis is directed at determining whether transgenes from GM virus-protected plants can move to other plants. Here again, for a proper risk assessment one should determine whether the gene flow occurs between the crop and its wild relative, whether the wild relatives are susceptible to the virus, and whether the virus infection would convey any fitness advantage to the plant (Tepfer, 2002). Concern about the breakdown of resistance to the virus within the GM plant can also be a concern, but whether this is a question of biosafety or just loss of efficacy of the plant is not clear. A recent study on transgenic papaya examined the possibility of recombination between transgenes and attacking PRSV, thus reducing the effectiveness of the GM papaya. In this case the coat protein gene of a mild strain of the PRSV caused posttranscriptional gene silencing, and the results indicate that PRSV recombinants posed minimal risk to commercial plantings of GM papaya (Kosiyachinda, 2002).

ENVIRONMENTAL CONSEQUENCES OF TWO NOVEL GM ORGANISMS

As other transgenic organisms are being developed and tested, it is important to consider their potential impact on the environment. The USDA is considering the ecological consequences and gene containment requirements of many different organisms, including grasses, trees, insects, and shellfish. Two examples, containment of transgenic fish and medically important proteins produced in plants are in the forefront of this new area.

Containment of Fish Genes

The risk of releasing a transgenic organism into the environment can be assessed by examining the probability of exposure to the hazard, which is equal to the expected long-term outcome of natural selection for the transgene if it escaped into the environment. Although escape will result in initial exposure, any harm will result from long-term exposure by the transgene (Muir, 2002). Over time, the transgene can become fixed in the population or can be eliminated through natural selection. Thus, the key component in understanding the risk is developing an understanding of the fitness costs that may be incurred. For example, in the case of transgenic fish that grow faster, Muir (2002) uses a number of factors, including adult survival, male fertility, relative mating success, and several other key factors to develop a model assessing fitness components. Muir and Howard (2001) developed a model which concluded that if 60 transgenic fish were introduced into a population of 60,000, the transgene would become fixed in the population in more than 40 generations. Although it is not certain what ecological consequences would arise from this complex situation, caution should be exercised.

Crops Bred to Produce Medicines

Crops can be genetically engineered to produce medicines, and industry officials hope that bio-farming will grow into a multibillion-dollar business by the end of the decade (Kilman, 2002). One promising method is to use plants to produce antibodies that can defend against diseases or be used in organ transplants, cancer therapy, and prevention of diseases. At present, antibodies used for medicines are produced in mammalian cell cultures in the laboratory, which has limited their availability and increased their costs. However, plants can serve as "production sites" for antibodies, which are later extracted from the plant and used in treatments. Another example is the production of plants that, when eaten, will produce immune responses. Potatoes and bananas have been developed through biotechnology to act like vaccines to prevent such diseases as traveler's diarrhea.

Although the current production of medicine-producing corn is less than 100 acres, considerable debate on the issue has already begun. In general, the food industry in the United States is supportive of GM crops but has concerns about the use of medicine-producing crops because the modified traits could possibly move into the food system through pollen drift or mishandling of the crop. The biotech industry has responded that its system would be a "closed system" in which the crop is grown in a location away from food crops, and that harvesting and transport of the product would be done by specialized vehicles used only for these GM crops. The biotech industry has used corn as the main crop for producing medicines because of its capacity to produce larger amounts of medically useful proteins in the kernels, but the food industry is urging that nonfood crops, such as tobacco, be used. A recent report (The Associated Press, 2002) stated that the government has ordered a biotechnology company to destroy 500,000 bushels of soybeans because they were contaminated with GM volunteer corn from the previous year's crop—corn that was grown to produce antibodies for medical use. A routine inspection by the FDA caught this infraction by ProdiGene, Inc., so there was no risk to the public, but it again points out the difficulty of regulating beneficial crops.

PESTICIDE-USE PATTERNS

In 2000, the estimated market for herbicides and insecticides was $13.7 billion and $8.0 billion, respectively (James, 2000). Since the present GM crops have been used for insect and weed management, an analysis of their effect on pesticide-use patterns is appropriate.

Bt Cotton

Cotton received the most insecticide of any crop worldwide, which is one of the reasons it was targeted as one of the first Bt crops. The National Center for Food and Agricultural Policy (NCFAP) conducted an analysis of the influence of Bt cotton on insecticide-use patterns (Gianessi and Carpenter, 1999; Carpenter and Gianessi, 2001). Insecticide use was compared for 1995, the year before Bt cotton varieties were introduced, and for 1998 and 1999, with adjustments for differences in area planted for these two years. Using data from six states, the results indicate that an overall reduction occurred in the use of insecticides for the bollworm/budworm complex and pink bollworm of >2 million pounds in 1998 and 2.7 million pounds in 1999. The number of insecticide applications also declined by 8.7 million in 1998 and 15 million in 1999, or 13 percent and 22 percent of the total number of insecticide applications in 1995. The authors noted that some of this reduction may have been due to other factors, such as the simultaneous boll weevil eradication programs that allowed beneficial insects to increase and control lepidoptera. On the other hand, they also noted that because Bt controls only lepidoptera, the number of secondary pests such as tarnished plant bug and stink bugs may have increased, and insecticides may have been targeted against them in Bt cotton fields. The EPA (2000) estimates that in 1999 a 7.5-million-acre treatment reduction took place when the figure is applied to the 13.3 million acres planted that year. An analysis by Williams (1999) of insecticide use in six states for control of the lepidopteran complex also indicates substantial reductions owing to the use of Bt cotton. In 1995, prior to the introduction of Bt cotton, the number of insecticide treatments ranged from 2.9 (Arizona) to 6.7 (Alabama), and averaged 4.8. By 1998 the range varied from 3.5 (Louisiana) to 0.4 (Arizona), and averaged 1.9, an overall reduction of 60 percent. In 1997 the use of Bt cotton in Arizona for pink bollworm eliminated 5.4 insecticide applications and saved growers $80 per acre (Carriere, 2001). In China, Bt cotton plants have contributed to a 60 to 80 percent decrease in the use of foliar insecticides (Xia et al., 1999). This study also estimated a reduction of 15,000 tons of pesticide. In a survey of 283 cotton farmers in China, Pray and colleagues (2001) reported that farmers "using Bt cotton reported less pesticide poisonings than those using conventional cotton." In Australia from 1996 to 2000, Bt cotton reduced insecticide use for bollworms by 4.1 to 5.4 sprays (43 to 57 percent) per year, with an overall reduction of all sprays from 37 to 52 percent (cited in Shelton et al., 2002). In Mexico, the use of Bt cotton allowed growers to save 55,090 liters of foliar insecticides and 4613 pesos/ha (Morales, 2000).

Bt Corn

In the United States and Canada, ECB is the most damaging lepidopteran insect, and losses resulting from its damage and control costs exceed $1 billion yearly. In a four-year study in Iowa, the combined losses due to first- and second-generation borers were 25 bushels per acre (Ostlie et al., 1997). ECB populations vary considerably by region and by year, and management practices are tailored accordingly. However, on a national scale "more farmers ignored ECB than treat it with insecticides" (Carpenter and Gianessi, 2001). NCFAP reports that approximately 1.5 million pounds of active ingredient were used to control ECB in 1996, nearly 60 percent of which was the organophosphate insecticide chlorpyrifos (Gianessi and Carpenter, 1999). In 1997, 7.1 percent (5.9 million acres) of the total corn planting in the United States was treated with insecticides for ECB, but in some regions this figure reached 25 percent.

Most corn growers do not use insecticides to control ECB because foliar sprays do not provide adequate control, so they accept the losses ECB causes. However, with the advent of Bt corn, growers can now get excellent control, far better than what was possible previously. This has led some to imply that Bt corn is not needed since growers did not treat their corn previously (Obrycki et al., 2001). A more reasoned approach to this question of whether growers need Bt corn was stated by Ortman and colleagues (2001) who pointed out that Bt corn was needed in certain areas where ECB was abundant and would cause economic damage. In 1997, 30 percent of the growers who planted Bt corn indicated they did so to eliminate the use of foliar insecticides for control of ECB (Carpenter and Gianessi, 2001). Growers' buying habits seem to validate these statements because the percentage of Bt corn in the total crop grew from <1 percent in 1996 to 26 percent in 2001. Comparing 1995, the year before Bt corn was introduced, to 1999, the use of five recommended insecticides for control of ECB declined. Carpenter and Gianessi (2001) concluded that a 1.5 percent decline in their use was due to Bt corn, amounting to approximately 1 million acres not sprayed for ECB control. A survey of Bt corn producers ($n = 7,265$) from six states (Illinois, Iowa, Kansas, Minnesota, Nebraska, and Pennsylvania) after each of the first three growing seasons when Bt corn was available for commercial production, documents that insecticide use for ECB is declining. The percentage of Bt corn producers that used less insecticide for this pest nearly doubled from 13.2 to 26.0 percent during the three-year period (Hellmich et al., 2000).

The EPA also did an analysis of the impact of Bt corn in six states for which statistics for annual insecticide use on corn were available for 1991-1999. The states were divided into high adopters (>25 percent of corn is Bt) (Iowa, Illinois, Nebraska, Missouri) and low adopters (<10 percent of corn is Bt) (Indiana and Wisconsin). In the high-adoption states, the use of insecticides decreased "from 6.0 million to slightly over 4 million acre treatments in 1999, a reduction of about one-third" (EPA, 2000). No such decline was observed for low-adopter states.

Different Philosophies on Bt Plants

In another review of the data, Benbrook (2001) agreed that Bt cotton "has reduced insecticide use in several states," but also agreed with other reports that the effect of Bt corn on insecticide use has been more complex and less dramatic. However, it must again be recognized that prior to the introduction of Bt corn, growers did not have a reliable method of controlling insects, and therefore accepted the reduced yield. Benbrook (2000) suggests that Bt crops perpetuate heavy reliance on treatments and continue the "pesticide treadmill" but does not offer any reliable alternatives of how to prevent injury and economic losses (Gianessi, 2000). Meanwhile, it has been estimated that of the $8.1 billion spent annually on all insecticides worldwide, nearly $2.7 billion could be substituted with Bt biotechnology products (Krattiger, 1997).

Herbicide-Tolerant Crops

The primary reason growers have adopted herbicide-tolerant crops is ease of weed control, and this has also led to changes in herbicide-use patterns. Citing USDA statistics, Carpenter (2001) noted that the number of pounds of herbicides used per soybean acre was the same in 1999 as in 1995, the year before HT soybeans were introduced. However, glyphosate was used on 20 percent of the area in 1995 and 62 percent in 1999. Growers decreased the number of herbicide applications by 12 percent, indicating they used fewer active ingredients and fewer trips to the field for weed control. In soybeans, the use of glyphosate, a class IV "practically nontoxic" herbicide, essentially replaced the other, more toxic classes. Benbrook (2001), a frequent critic of biotechnology, largely agrees that HT varieties have reduced the number of active ingredients applied per area but suggests that HT crops have "modestly increased the average pounds applied per acre." He does not dispute the increased safety of glyphosate over the alternative herbicides.

In Argentina, which has the second largest transgenic area in the world, HT soybeans are grown widely because there are no wild relatives. In an analysis, Qaim and Traxler (2002) found that glyphosate substitutes for a number of other products, resulting in a decrease in cost. However, the average number of applications increased slightly and the amounts per ha went up considerably. This contrasted with reports in the United States in which the number of applications went down and the aggregate herbicide amount stayed the same. However, an important point is not the total volume of pesticide used, but the types of pesticides employed. Herbicides are divided into classes based on their toxicity to humans. In Argentina, the use of glyphosate led to the complete elimination of herbicides belonging to the higher toxicity classes II and III—there are no class I herbicides (Qaim and Traxler, 2002). In addition to the pesticide aspects, nearly double the number of growers who used Roundup Ready (RR) soybeans used no-till practices, which eliminated one tillage operation and resulted in a fuel savings of 10 liters/ha. It also helped preserve soil texture and wind and water erosion. In Canada, a recent study indicated that canola growers planting HT varieties eliminated 6,000 metric tons of herbicide (Canola Council of Canada, 2001).

RESISTANCE MANAGEMENT TO HERBICIDE-RESISTANT AND INSECT-RESISTANT PLANTS

Over time, some insects, weeds, and pathogens have evolved resistance to specific pesticides, and it hasn't mattered whether the pesticides occur naturally or have been synthesized. Pesticide resistance is an evolutionary process caused by a genetic change in the pest in response to selection pressure, resulting in strains of the pest capable of surviving a dose lethal to a majority of individuals in a population. There is justified concern that plants expressing pesticidal properties may, over time, lose some of their effectiveness due to resistance.

More than 109 weed biotypes are resistant to herbicides, and over half of the cases involve one class—the triazines (Lebanon, 1991). After 25 years of intensive use, only three species of weeds have been documented as becoming resistant to glyphosate (Carpenter et al., 2002). However, this may change with increased use of glyphosate-tolerant crops. Diversifying weed management through the use of rotational patterns, mechanical cultivation, or different herbicides will lessen the selection pressure for glyphosate-resistant weeds. Unlike

Bt-insect-protected plants, HT crops have no mandatory resistance management programs.

The evolution of insects' resistance to the toxins produced by Bt plants depends on the genetic basis of resistance, the initial frequency of resistance alleles in the population, the competitiveness of resistant individuals in the field, and a resistance management strategy (Shelton et al., 2002). The EPA mandates a resistance management strategy for the use of Bt crops; this is based on the plants expressing a high dose of the toxin and the use of a refuge to conserve susceptible alleles within the population. Modeling studies and greenhouse and field experiments have shown the effectiveness of the high-dose refuge strategy (Shelton et al., 2002), but care must be taken in managing the insect population within the refuge to ensure that sufficient susceptible alleles will exist. After six years of extensive use of Bt crops in the field, no cases have been reported of field failure due to resistance. However, because there have been cases of insects developing resistance to foliar sprays of Bt (Tabashnik, 1994), this should remain a concern.

The usefulness of trying to detect shifts in the frequency of resistance genes prior to insects beginning to survive on the plants in the field and the question of whether monitoring can be done through molecular methods has engendered considerable debate. The EPA does require a monitoring program and a mitigation strategy if resistance occurs (EPA, 2000). While the first generation of insect-protected Bt plants are in use, efforts should also be undertaken to develop alternatives to Bt plants, or Bt plants that express multiple toxins or express toxins in a manner to further delay the onset of resistance. Cotton that expresses two Bt toxins has been registered in Australia and is soon expected to be registered in the United States. Although some may see this as a continuation of the pesticide treadmill, research has shown that the deployment of plants with pyramided Bt genes will greatly delay the onset of resistance and lead to prolonged use of a technology that has shown positive environmental and human health benefits in cotton. In the future, plants may also be engineered to express the Bt toxins (or other toxins) only at specific times, and this should lead to increased delays in the ability of insects to develop resistance (Shelton et al., 2002).

AGRICULTURAL BIOTECHNOLOGY
IN A CHANGING WORLD

Much has been promised by biotechnology, and some products have been delivered. As with most new technologies, the benefits of biotechnology in the short term may have been overstated by its proponents. However, science is only beginning to glimpse the long-term effect biotechnology will have as it becomes woven into products that affect our daily lives. Many common drugs, including the majority of insulin used today, are produced through biotechnology. GM enzymes play an increasingly important role in a variety of processes to manufacture fuels, detergents, textiles, and wood and paper products. These applications have not experienced the level of controversy surrounding the present and perhaps future applications of agricultural biotechnology. To some extent, this is due to consumers not seeing the direct benefits of the present GM crops developed for pest-management purposes. GM plants that will have increased nutrition aspects (e.g., better oil profiles or higher vitamin content) may enhance public acceptance of GM plants, although these plants will also have some of the same environmental and health concerns as the present GM plants. However, if the public sees more benefits in these consumer-oriented plants, it may judge the potential benefits to outweigh the potential risks, and public support may increase.

The development of Golden Rice may be an example of this. Golden Rice contains increased levels of beta-carotene and other carotenoids needed for the production of vitamin A, essential in the prevention of blindness. Although not even the developers of this rice believe it to be a silver bullet in the fight against malnutrition, they do believe it to be a component in the overall solution. Likewise, biotechnology is often seen as the means to help feed the ever-expanding population of the world by providing plants that can grow in previously unsuitable habitats, provide their own fertilizers, or provide enhanced yield and nutrition. Agricultural biotechnology is, again, only one component of an overall system needed to help feed the world, and it must be accompanied by improved distribution of food to those who need it most.

Regardless of who benefits from products derived from GM plants and animals, social and legal questions will continue to surface. Many of these questions focus on who develops and controls the technologies and who will pay for damages if something goes awry. If organic growers' crops are contaminated by pollen from GM crops, who will pay for their losses? Likewise, if a conventional grower is not able to use Bt corn because of the potential for pollen drift into his neighbor's organic corn, what options does he have for control, and will these options have a broader negative impact on the environment? Another frequently posed question deals with organic growers losing one of their tools if resistance to Bt plants occurs. From an overall public good perspective, should the limited use of Bt sprays by organic growers outweigh the broader good derived from its more extensive use in Bt plants? These are difficult legal and social questions affecting the deployment of GM crops and often seem to pit organic agriculture against biotechnology. This is not helpful because aspects of both types of agriculture can be used to reduce the negative impact of agriculture on the environment.

As with most new technologies, biotechnology was developed from partnerships with the public and private sectors. Amid increasingly expensive research technologies and simultaneously declining federal and state support, university scientists have had to search for funding outside the shrinking pool of public support. The result has been even closer partnerships between the private and public sectors, and this has raised ethical questions. Scientists that are funded by the public have an obligation to work for the public good. To do so now may require them to work more closely with the private sector to ensure their discoveries make it to the marketplace to produce the intended public good. At the same time, however, one must ask whether universities presently devote too much effort to biotechnology at the expense of other programs. Again, this question must be addressed through a risk-benefit analysis between the many choices for agricultural research programs.

Other ethical questions arise, including whether a downside exists to the growing presence of corporate money and private investment in the relationship among science, technology, universities, and products. If so, what other options are available and what risks and benefits do they offer? One concern is the possibility that the exchange of ideas among scientists, the public, and the private sector is more limited because of patents and private intellectual rights. This potential risk must be balanced against the potential benefit of public good that comes from a new, useful technology being developed because of the partnership between public and private goals.

It is important for the public to become engaged in an informed dialogue about the risks and benefits of agricultural biotechnology, but this has been difficult because of the often polarized way in which biotechnology is presented. It is often seen as part of an overall trend toward globalization, reduction in small farms, and other problems with the fabric of our complex society, but it should be clear that these trends began long before the advent of modern biotechnology. Those who support and those who oppose agricultural biotechnology should have the same goal—production of a safe, abundant, tasty, economical, diverse, and nutritious food supply that is grown in an environmentally and socially responsible fashion. To the extent that any technology can bring us closer to this goal, it should be supported.

REFERENCES

Abelson, P.H. (1998). A third technological revolution. *Science* 279:2019.

Appell, D. (2001). The new uncertainty principle. *Sci. Am.* January:18-19.

The Associated Press (2002). FDA orders destruction of soybeans contaminated with genetically engineered corn. November 12, newswire.

Barkley, A.P. (2002). The economic impacts of agricultural biotechnology on international trade, consumers, and producers: The case of corn and soybeans in the USA. *Agricultural Biotechnologies: New Avenues for Production, Consumption, and Technology Transfer,* Sixth International Consortium on Agricultural Biotechnology and Research (ICABR) Conference, Ravello, Italy, July.

Beachy, R., Bennetzen, J.L., Chassy, B.M., Chrispeels, M., Chory, J., Ecker, J., Noel, J., Kay, S., Dean, C., Lamb, C., et al. (2002). Divergent perspectives on GM food. *Nat. Biotechnol.* 20:1195-1196.

Beever, D.E. and Kemp, C.F. (2000). Safety issues associated with the DNA in animal feed derived from genetically modified crops. A review of the scientific and regulatory procedures. *Nutr. Abstr. Rev.* 70(3):175-182.

Benbrook, C. (2000). An appraisal of EPA's assessment of the benefits of *Bt* crops: Prepared for the Union of Concerned Scientists. October 17.

Benbrook, C. (2001). Do GM crops mean less pesticide use? *Pesticide Outlook* October:204-207.

Berg, P., Baltimore, D., Boyer, H.W., Cohen, S.N., Davis, R.W., Hogness, D.S., Nathans, D., Roblin, R., Watson, J., Weissman, S., and Zinder, N. (1974). Potential biohazards of recombinant DNA molecules. *Science* 185(4184):303.

Berg, P., Baltimore, D., Brenner, S., Roblin, R.O., and Singer, M. (1975). Asilomar conference on recombinant DNA molecules. *Science* 188(4192):991-994.

Betz, F.S., Hammond, B.G., and Fuchs, R.L. (2000). Safety and advantages of *Bacillus thuringiensis*-protected plants to control insect pests. *Reg. Toxicol. Pharmacol.* 32:156-173.

Brown, C., Campbell, I., and Priest. F. (1987). *Introduction to Biotechnology.* Blackwell Scientific Publications, Oxford, England.

Canola Council of Canada (2001). An agronomic and economic assessment of transgenic canola. <http://www.canola-council.org>, January.

Carpenter, J.E. (2001). Case studies in benefits and risks of agricultural biotechnology: Roundup Ready soybeans and *Bt* field corn. National Center for Food and Agriculture Policy, <http://www.ncfap.org/pubs.htm#Biotechnology>.

Carpenter, J., Felsot, A., Goode, T., Haming, M., Onstad, D., and Sankula, S. (2002). *Comparative Environmental Impacts of Biotechnology-Derived and Traditional Soybean, Corn, and Cotton Crops.* Council for Agricultural Science and Technology, Ames, IA.

Carpenter, J.E. and Gianessi, L.P. (2001). Agricultural biotechnology: Updated benefit estimates. National Center for Food and Agriculture Policy, <http://www.ncfap.org/pubs.htm#Biotechnology>.

Carriere, Y., Dennehy, T.J., Pedersen, B., Haller, S., Ellers-Kirk, C., Antilla, L., Liu, Y.-B., Willot, E., and Tabashnik, B.E. (2001). Large-scale management of insect resistance to transgenic cotton in Arizona: Can transgenic insecticidal crops be sustained? *J. Econ. Entomol.* 94:315-325.

Centers for Disease Control (CDC) (2001). *Investigation of Human Health Effects Associated with Potential Exposure to Genetically Modified Corn: A Report to the US Food and Drug Administration from Centers for Disease Control and Prevention.* <http://www.cdc.gov/nieh/ehhe/Cry9cReport/Cry9report.pdf>.

Chapman, B., Morris, S., and Powell, D. (2002). 2001 world review of agbiotechnology. *ISB News Report* January:1-6.

Chassy, B.M. (2002). Food safety assessment of current and future plant biotechnology products. In *Biotechnology and Safety Assessment* (pp. 87-115), J.A. Thomas and R.L. Fuchs, eds. Academic Press, San Diego, CA.

Daniell, H. (2002). Molecular strategies for gene containment in transgenic crops. *Nat. Biotechnol.* 20(6):581-586.

Ellstrand, N.C., Prentice, H.C., and Hancock, J.F. (1999) Gene flow and introgression from domesticated plants into their wild relatives. *Ann. Rev. Ecol. Syst.* 30: 539-563.

Environmental Protection Agency (EPA) (1995). *Pesticide Fact Sheet for* Bacillus thuringiensis *subsp.* kurstaki CryI(A)b *Delta-Endotoxin and the Genetic Material Necessary for the Production (Plasmid Vector pCIB4431) in Corn,* Publ. EPA731—F-95—004, EPA, Washington, DC.

Environmental Protection Agency (EPA) (1999). *Cry9C Food Allergenicity Assessment Background Document.* <http://www.epa.gov/oppbppd1/biopesticides/pips/old/cry9c/cry9c-epa-background.htm>.

Environmental Protection Agency (EPA) (2000). *Biopesticides Registration Document: Preliminary Risks and Benefits Sections;* Bacillus thuringiensis *Plant-pesticides.* EPA, Washington, DC.

Environmental Protection Agency (EPA) (2001a). *Bt Cotton Refuge Requirements for the 2001 Growing Season.* <http://www.epa.gov/pesticides/biopesticides/pips/bt_cotton_refuge_2001.htm>.

Environmental Protection Agency (EPA) (2001b). *Bt Plant-Pesticides Risk and Benefit Assessments, 2000 FIFRA SAP Report #2000-07.* <http:// www.epa.gov/scipoly/sap/2000/index.htm#october>.

European Commission, Health and Consumer Protection Directorate-General (2002). Opinion of the Scientific Committee on Food on the safety assessment of the genetically modified maize line GA21, with tolerance to the herbicide glyphosate. March 6. <http://europa.eu.int/comm/food/fs/sc/scf/out121_en.pdf>.

Falck-Zepeda, J., Traxler, G., and Nelson, R.G. (2000). Surplus distribution from the introduction of a biotechnology innovation. *Am. J. Agric. Econ.* 82:360-369.

Faust, M.A. (2000). Straight talk about biotech crops for the dairy industry. Invited presentation for the 2000 World Dairy Expo, Madison, WI, October 7.

Federation of Animal Science Societies (2001). Communications. References: Feeding transgenic crops to livestock. <http://www.fass.org/REFERENC.htm>.

Folmer, J.D., Grant, R.J., Milton, C.T., Beck, J.F. (2000). Effect of *Bt* corn silage on short-term lactational performance and ruminal fermentation in dairy cows. *J. Dairy Sci.* 83:1182 (Abstr. 272).

Food and Agriculture Organization (FAO) (2003). Codex Alimentarius Commission adopts more than 50 new food standards. *FAO Newsroom,* <http://www.fao.org/english/newsroom/news/2003/20363-en.html>.

Food Standards Australia New Zealand (2002). Food standards code. <http://www.foodstandards.gov.au/foodstandardscode/>.

Forrester, N. (1997). The benefits of insect resistant cotton. In *Commercialisation of Transgenic Crops: Risk, Benefit, and Trade Considerations* (pp. 239-242), G.D. McLean, P. Waterhouse, G. Evans, and M.J. Gibbs, eds. Cooperative Research Centre for Plant Science and Bureau of Statistics, Canberra, Australia.

Fresh Trends (2001). Profile of the Fresh Produce Consumer. The Packer, Vance Publ. 2001.

Frisvold, G.B., Tronstad, R., and Mortensen, J. (2000). Adoption of *Bt* cotton: Regional differences in producer costs and returns. *Proc. Beltwide Cotton Conf.* 2:337-240.

Frutos, R., Rang, C., and Royer, M. (1999). Managing resistance to plants producing *Bacillus thuringiensis* toxins. *Crit. Rev. Biotechnol.* 19:227-276.

Gaisford, J.D. and Chui-Ha, C.L. (2000). The case for and against import embargoes on products of biotechnology. *Estey Ctr. J. Intl. Law Trade Policy* 1(1):83-98. <http://www.cafri.usask.ca/estey>.

Gianessi, L.P. (2000). A critique of: An appraisal of EPA's assessment of the benefits of *Bt* crops by C. Benbrook. National Center for Food and Agricultural Policy, October 31.

Gianessi, L.P. and Carpenter, J.E. (1999). *Agricultural Biotechnology: Insect Control Benefits*. National Center for Food and Agricultural Policy, Washington, DC.

Gill, S.S., Cowles, E.A., and Pietrantonio, P.V. (1992). The mode of action of *Bacillus thuringiensis* endotoxins. *Ann. Rev. Entomol.* 37:615-636.

Goklany, I. (2002). From precautionary principle to risk-risk analysis. *Nat. Biotechnol.* 20(11):1075.

Haney, R.L., Senseman, S.A., Hons, F.M., and Zuberer, D.A. (2000). Effect of glyphosate on soil microbial activity and biomass. *Weed Sci.* 48:89-93.

Hellmich, R.L., Rice, M.E., Pleasants, J.M., and Lam, W.K. (2000). Of monarchs and men: Possible influences of *Bt* corn in the agricultural community. *Proceedings of the Integrated Crop Management Conference* (pp. 85-94), Iowa State University Extension, Iowa State University, Ames.

Hilbeck, A., Baumgartner, M., Freid, P.M., and Bigler, F. (1998). Effects of *Bacillus thuringiensis* corn-fed prey on mortality and development time of immature *Chrysoperla carnae*. *Environ. Entomol.* 27:480-487.

Hobbs, J.E., Gaisford, J.D., Isaac, G.E., Kerr, W.A., and Klein, K.K. (2002). International commercial policy toward agricultural biotechnology: Issues and evolution. *Agricultural Biotechnologies: New Avenues for Production, Consumption, and Technology Transfer*, Sixth International ICABR Conference, Ravello, Italy, July.

Huffman, W.E., Rousu, M., Shogren, J.F., and Tegene, A. (2002). Should the United States initiate a mandatory labeling policy for genetically modified foods? *Agricultural Biotechnologies: New Avenues for Production, Consumption, and Technology Transfer*, Sixth International ICABR Conference, Ravello, Italy, July.

Institute of Food Technologists (2000). Expert report on biotechnology and foods: Human food safety evaluations of rDNA biotechnology-derived foods. <http://www.ift.org/>.

International Food Information Council (IFIC) (2002). Survey results. FoodBiotech Net, Council for Agricultural Science and Technology (CAST), September 23.

James, C. (2000). Global status of commercialized transgenic crops: 2000. *ISAAA Briefs* No. 21, International Service for the Acquisition of Agri-Biotech Applications (ISAAA), Ithaca, NY.

James, C. (2001a). Global status of commercialized transgenic crops: 2000. *ISAAA Briefs* No. 23, ISAAA, Ithaca, NY.

James, C. (2001b). Global status of commercialized transgenic crops: 2001. *ISAAA Briefs* No. 24, ISAAA, Ithaca, NY.

Janinski, J., Eisley, B., Young, C., Wilson, H., and Kovach, J. (2001). *Beneficial Arthropod Survey in Transgenic and Non-Transgenic Field Crops in Ohio* (pp. 1-2). Ohio State University: Ohio Agricultural Research and Development Center, Columbus.

Jia, S.R., Guo, S.D., and An, D.C., eds. (2001). *Transgenic Cotton*. Science Press, Beijing.

Jones, P.B.C. (2002) Litigation in the wind. *ISB News Report* April:10-11.

Kelves, D.J. (2001). The battle over biotechnology. In *Days of Destiny: Crossroads in American History* (pp. 453-463), D. Rubel, ed. DK Publishing, New York.

Kilman, S. (2002). Crops bred to produce medicines raise contamination worries. *The Wall Street Journal*, November 5.

Kirsten, J. and Gouse, M. (2002). *Bt* cotton in South Africa: Adoption and impact on farm incomes amongst small- and large-scale farmers. *ISB News Report* October:7-9.

Kosiyachinda, P. (2002). Risk assessment of recombination between transgene and attacking virus by using virus-resistant transgenic papaya as a model system. PhD thesis, Cornell University.

Krattiger, A.F. (1997). Insect resistance in crops: A case study of *Bacillus thuringiensis (Bt)* and its transfer to developing countries. *ISAAA Briefs* No. 2, ISAAA, Ithaca, New York.

Lebanon, H.M. (1991). Herbicide-resistant weed continue to spread. *Res. Pest Mgmt. Newsl.* 3:36-37.

Lin, W. (2002). Segregation of non-biotech corn and soybeans: Who bears the cost? *Agricultural Biotechnologies: New Avenues for Production, Consumption and Technology Transfer,* Sixth International ICABR Conference, Ravello, Italy, July.

Losey, J., Raynor, L., and Carter, M.E. (1999). Transgenic pollen harms Monarch larvae. *Nature* 399:214.

Marvier, M. (2001). Ecology of transgenic crops. *Am. Sci.* 89:160-167.

Metcalf, R.L. and Metcalf, R.A. (1993). *Destructive and Useful Insects: Their Habits and Control,* Fifth Edition. McGraw-Hill, New York.

Morales, A.A. (2000). *The Mexican Experience with Insect Resistant* Bt *Transgenic Crops.* (Abstr.). Presented to the Society of Invertebrate Pathology, Guanajuato, Mexico, August 13-18.

Muir, W.M. (2002). Potential environmental risks and hazards of biotechnology. Part II: Methods to estimate risks and hazards. *ISB News Report* February:1-4.

Muir, W.M. and Howard, R.D. (2001). Fitness components and ecological risk of transgenic release: A model using Japanese medaka *(Oryzias latipes). Am. Naturalist* 158:1-16.

Munkvold, G.P. and Desjardins, A.E. (1997). Fumonisins in maize: Can we reduce their occurrence? *Plant Dis.* 81:556-565.

Munkvold, G.P., Hellmich, R.L., and Rice, L.G. (1999). Comparison of fumonisin concentrations in kernels of transgenic *Bt* maize hybrids and non-transgenic hybrids. *Plant Dis.* 83:130-138.

National Academy of Sciences (NAS) (2000). *Genetically Modified Pest-Protected Plants: Science and Regulation.* National Academy Press, Washington, DC.

National Academy of Sciences (NAS) (2002). *Environmental Effects of Transgenic Plants: The Scope and Adequacy of Regulation.* National Academy Press, Washington, DC.

Nature (2002). Editorial note. *Nature* 416:603.

Nickson, T.E. and McKee, M.J. (2002). Ecological assessment of crops derived through biotechnology. In *Biotechnology and Safety Assessment* (pp. 233-251), J.A. Thomas and R.L. Fuchs, eds. Academic Press, San Diego, CA.

Obrycki, J.J., Losey, J.E., Taylor, O.R., and Jesse, L.C.H. (2001). Transgenic insecticidal corn: Beyond insectidal toxicity to ecological complexity. *BioScience* 51: 353-361.

Ortman, E.E., Barry, B.D., Bushman, L.L., Calvin, D.D., Carpenter, J., Dively, G.P., Foster, J.E., Fuller, B.W., Hellmich, R.L., Higgins, R.A., et al. (2001). Transgenic insecticidal corn: The agronomic and ecological rationale for its use. *Bioscience* 51:900-903.

Ostlie, K.R., Hutchinson, W.D., and Hellmich, R.L. (1997). *Bt* corn and European corn borer: Long-term success through resistance management. *North Central Regional Extension Publication 602*. University of Minnesota Extension Service, St. Paul, MN.

Pachico, D. and Wolf, M. (2002). Attitudes toward genetically modified food in Columbia. *Agricultural Biotechnologies: New Avenues for Production, Consumption, and Technology Transfer*, Sixth International ICABR Conference, Ravello, Italy, July.

Padgette, S.R., Re, D.B., Barry, G.F., Eichholtz, D.E., Delannay, X., Fuchs, R.L., Kishore, G.M., and Fraley, R.T. (1996). New weed control opportunities: Development of soybeans with a Roundup Ready gene. In *Herbicide Resistant Crops—Agricultural, Enviornmental, Economic, Regulatory, and Technical Aspects* (pp. 53-84), S.O. Duke, ed. CRC Press, Boca Raton, FL.

Pennings, J.M.E., Wansink, B., and Meulenberg, M.T.G. (2002) A note on modeling consumer reactions to a crisis: The case of the mad cow disease. *Intern. J. Marketing* 19:91-100.

Perdikis, N. (2000). A conflict of legitimate concerns or pandering to vested interests? Conflicting attitudes towards the regulation of trade in genetically modified goods—The EU and the US. *Estey Ctr. J. Intl. Law Trade Policy*, 1(1):51-65. <http://www.cafri.usask.ca/estey>.

Pray, C.E., Ma, D., Huang, J., and Qiao, F. (2001). Impact of *Bt* cotton in China. *World Dev.* 29:813-825.

Qaim, M. and Traxler, G. (2002). Roundup Ready soybeans in Argentina: Farm level, environmental, and welfare effects. *Agricultural Biotechnologies: New Avenues for Production, Consumption, and Technology Transfer*, Sixth International ICABR Conference, Ravello, Italy, July.

Quist, D. and Chapela, I.H. (2001). Transgenic DNA introgressed into traditional maize landraces in Oaxaca, Mexico. *Nature* 414:541-543.

Reiss, M. (2002). Labeling GM foods—The ethical way forward. *Nat. Biotechnol.* 20(9):868.

Saxena, D., Flores, S., and Stotzsky, G. (1999). Insecticidal toxin in root exudates from *Bt* corn. *Nature* 402:480.

Saxena, D. and Stotzsky, G. (2001). Fate and effects of the insecticidal toxins from *Bacillus thuringiensis* in soil. *Inf. Syst. Biotechnol. News Rep.* May 5-8.

Sears, M.K., Hellmich, R.L., Stanley-Horn, D.E., Oberhauser, K.S., Pleasants, J.M., Mattila, H.R., Siegfried, B.D., and Dively, G.P. (2001). Impact of *Bt* corn pollen on monarch butterfly populations: A risk assessment. *Proc. Natl. Acad. Sci.* (USA) 98:11937-11942.

Shelton, A.M. and Sears, M.K. (2001). The monarch butterfly controversy: Scientific interpretations of a phenomenon. *The Plant* 27:483-488.

Shelton, A.M., Zhao, J.-Z. and Roush, R.T. (2002). Economic, ecological, food safety, and social consequences of the deployment of *Bt* transgenic plants. *Ann. Rev. Entomol.* 47:845-881.

Shoemaker, R., Harwood, J., Day-Rubenstein, K., Dunahay, T., Heisey, P., Hoffman, L., Klotz-Ingram, C., Lin, W., Mitchell, L., McBride, W., and Fernandez-Cornejo, J. (2001). Economic Issues in Agricultural Biotechnology. *USDA/ERS Agricultural Bulletin No. 762*, March.

Smalla, K., Borin, S., Heuer, H., Gebhard, F., Dirk van Elsas, J., and Nielsen, K. (2000). Horizontal transfer of antibiotic resistance genes from transgenic plants to bacteria—Are there new data to fuel the debate? In *Proceedings of the Sixth International Symposium on the Biosafety of Genetically Modified Organisms*, C. Fairbairn, G. Scoles, and A. McHughen, eds. University Extension Press, Saskatchewan, Canada.

Smyth, S., Khachatourians, G.G. and Phillips, P.W.B. (2002). Liabilities and economics of transgenic crops. *Nat. Biotechnol.* 20(6):537-541.

Snow, A.A. (2002). Transgenic crops—Why gene flow matters. *Nat. Biotechnol.* 20(6):542.

Snow, A.A. and Palma, P.M. (1997). Commercialization of transgenic plants: Potential ecological risks. *BioScience* 47:86-96.

Society of Toxicology (2002). Position paper: The safety of genetically modified foods produced through biotechnology. <http://www.toxicology.org/Information/GovernmentMedia/GM_Food.html>, adopted September 25, 2002.

Tabashnik, B.E. (1994). Evolution of resistance to *Bacillus thuringiensis*. *Ann. Rev. Entomol.* 39:47-79.

Tepfer, M. (2002). Risk assessement of virus-resistant plants. *Ann. Rev. Phytopathol.* 40:467-491.

Traynor, P.L. and Westwood, J.H., eds. (1999). *Proceedings of the Workshop on Ecological Effects of Pest Resistance Genes Managed Ecosystems*, January 31 through February 3, Bethesda, Maryland. Blacksburg, VA: Information Systems Biotechnology <http://www.isb.vt.edu>.

U.S. Department of Agriculture, Animal and Plant Health Inspection Service (USDA APHIS) (2002). *Agricultural Biotechnology: Permitting, Notification, and Deregulation.* <http://www.aphis.usda.gov/brs/index.html>.

U.S. Department of Agriculture, National Agricultural Statistics Service (USDA NASS) (2001). *Acreage.* <http://usda.mannlib.cornell.edu/reports/nassr/field/pcp-bba/acrg2001.pdf>.

U.S. Department of Agriculture, National Agricultural Statistics Service (USDA NASS) (2002). *Prospective Plantings.* <http://usda.mannlib.cornell.edu/reports/ nassr/field/pcp-bbp/psp10302.pdf>.

Van der Sluis, E. and Van Scharrel, A. (2002). Farm level transgenic crop adoption rates in South Dakota. *ISB News Report* October:9-11.

The Washington Post, (2002). Editorial: Europe's biotech madness. *The Washington Post,* October 28:A18.

Watkinson, A.R., Freckleton, R.P., Robinson, R.A., and Sutherland, W.J. (2000). Predictions of biodiversity response to genetically modified herbicide-tolerant crops. *Science* 289:1554-1557.

Wilkinson, M. (2002). Gene flow from transgenic plants. In *Biotechnology and Safety Assessment* (pp. 413-433), J.A. Thomas and R.L. Fuchs, eds. Academic Press, San Diego, CA.

Willams, M.R. (1999). Cotton insect losses—1998 (Compiled for the National Cotton Council). <http://www.msstate.edu/Entomology/CTNLOSS/1998loss.html>.

Wolf, M.M., Bertolini, P., and Parker-Garcia, J. (2002). A comparison of consumer attitudes toward genetically modified food in Italy and the United States. *Agricultural Biotechnologies: New Avenues for Production, Consumption, and Technology Transfer,* Sixth International ICABR Conference, Ravello, Italy, July.

Wolfenbarger, L.L. and Phifer, P.R. (2001). GM crops and patterns of pesticide use. *Science* 292:637-638.

World Health Organization (WHO) (2000). *Safety Aspects of Genetically Modified Foods of Plant Origin.* Report of a joint FAO/WHO expert consulation on foods derived from biotechnology, Geneva, Switzerland, May 29 through June 2. <http:// www.who.int/fsf/GMfood/FAO-WHO_Consultation_report_2000.pdf>.

Xia, J.Y., Cui, J.J., Ma, L.H., Dong, S.X., and Cui, X.F. (1999). The role of transgenic *Bt* cotton in integrated pest management. *Acta Gossypii Sinica* 11:57-64.

Chapter 2

Sampling for the Detection of Biotech Grains: A USDA Perspective

Larry D. Freese

INTRODUCTION

A consignment of grain (or grain lot) has many unknown quality characteristics. Measuring these characteristics on the entire lot can be costly. An experienced inspector must examine the grain to determine whether a kernel is damaged or not. Time constraints and cost prohibit an inspector from examining every kernel in a grain lot. Inspecting a small subset of the lot is much less costly and time-consuming than inspecting the whole lot. This subset of the lot is called a sample.

Although inspecting a sample is much less costly than inspecting the entire lot, the content of a sample does not always reflect the content of the lot. Fortunately, when samples are properly taken, probability theory can assign some risk values to measurements on samples. Furthermore, sampling from a lot is only one source of error when estimating a quality characteristic of a lot. Sources of error fall into three basic categories: (1) sampling, (2) sample preparation, and (3) analytical method (Whitaker et al., 2001). Sampling is an ever-present source of error when estimating characteristics of a lot. However, depending on the characteristic being measured, sample preparation and analytical method can be significant contributors to measurement errors. Minimizing these errors is necessary to ensure better precision and accuracy in the final analytical result.

Buyers and sellers have to agree on the quality and price of the lot before a transaction can take place. Basing the quality of a lot on a sample introduces risk to both parties. Buyers and sellers want to control their risk where possible. Knowledge of the sources of error will help buyers and sellers make informed decisions.

This chapter discusses sampling errors associated with detecting the presence of biotech varieties in grain lots, and presents ways to estimate these errors and control the risks to buyers and sellers. The variability shown in the sample estimates assumes only sampling variability. No allowance for error from sample preparation or from analytical method has been incorporated into the estimate ranges. No specific sampling plan is recommended. Buyers and sellers should agree on a sampling and testing plan that best meets their mutual needs.

Three agencies of the U.S. government are primarily responsible for regulating biotechnology-derived products: Animal and Plant Health Inspection Service (APHIS) of the U.S. Department of Agriculture (USDA); the Environmental Protection Agency (EPA); and the Food and Drug Administration (FDA) of the the Department of Health and Human Services. An overview of responsibilities can be found online at a Web site of the U.S. Department of Agriculture, Animal and Plant Health Inspection Service (APHIS, 2002). The following are excerpts from this site:

> Within USDA, the Animal and Plant Health Inspection Service (APHIS) is responsible for protecting US agriculture from pests and diseases. Under the authority of the Federal Plant Pest Act, APHIS regulations provide procedures for obtaining a permit or for providing notification, prior to "introducing" a regulated article in the United States. Regulated articles are considered to be organisms and products altered or produced through genetic engineering that are plant pests of [*sic*] that there is reason to believe are plant pests. The act of introducing includes any movement into (import) or through (interstate) the United States, or release into the environment outside an area of physical confinement. The regulations also provide for a petition process for the determination of nonregulated status. Once a determination of nonregulated status has been made, the product (and its offspring) no longer requires APHIS review for movement or release in the US. (APHIS, 2002)

The EPA ensures the safety of pesticides, both chemical and those that are produced biologically. The BioPesticides and Pollution Prevention Division of the Office of Pesticide Programs (OPP) uses the authority of the Federal Insecticide, Fungicide, and Rodenticide Act (FIFRA) to regulate the distribution, sale, use, and testing of plants and microbes producing pesticidal substances (GAO, 2002). Under the Federal Food, Drug, and Cosmetic Act (FFDCA), the EPA sets tolerance limits for substances used as pesticides on and in food and feed, or establishes an exemption from the requirement of a tolerance (Ahmed, 1999). EPA also establishes tolerances for residues of herbicides used on novel herbicide-tolerant crops. Under authority of the Toxic Substances Control Act (TSCA), the EPA's TSCA Biotechnology Program regulates microorganisms intended for commercial use that contain or express new combinations of traits. This includes "intergeneric microorganisms" formed by deliberate combinations of genetic material from different taxonomic genera (GAO, 2002).

As a part of the Department of Health and Human Services, FDA regulates foods and feed derived from new plant varieties under the authority of the Federal Food, Drug, and Cosmetic Act. FDA policy is based on existing food law, and requires that genetically engineered foods meet the same rigorous safety standards as is required of all other foods. FDA's biotechnology policy treats substances intentionally added to food through genetic engineering as food additives if they are significantly different in structure, function, or amount than substances currently found in food. Many of the food crops currently being developed using biotechnology do not contain substances that are significantly different from those already in the diet and thus do not require pre-market approval. Consistent with its 1992 policy, FDA expects developers to consult with the agency on safety and regulatory questions. (FDA, 1992)

INTRODUCTION TO SAMPLING THEORY

A sample is simply a subset of a lot. Probability theory can describe risk for randomly selected samples. A simple random sample is one selected in a process in which every possible sample from a lot has an equal chance of being selected (McLeod, 1988). If every possible sample from a lot could be measured, the average of the measurements would equal the content of the lot. This means that, on average, a random sample produces an unbiased estimate of the measurement of interest. Measurements on individual samples will deviate from the content in the lot. Probability will not tell what the deviation is on a particular sample, but it can describe a likely range that the lot content will fall into. Suppose a random sample of 100 kernels is selected from a lot with 5 percent biotech kernels. The distribution for this example is given in Figure 2.1 and is based on the binomial distribution (Hogg and Craig, 1978). A sample from this lot would likely provide an estimate between one and nine percent biotech kernels (Figure 2.1). Increasing the sample size can reduce the range of estimated results. For example, a sample size of 800 kernels would provide an estimate between 3.4 and 6.6 percent biotech kernels (Figure 2.2).

Sampling from Grain Lots

In practice, a pure random sample is not always easy to obtain from a lot. A sampling technique called systematic sampling has been used widely to produce a sample that is a reasonable substitute for a random sample. For example, auditors may use systematic sam-

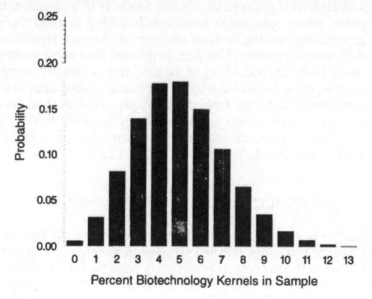

FIGURE 2.1. Distribution for 100-kernel samples

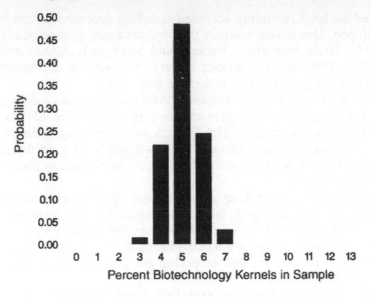

FIGURE 2.2. Distribution for 800-kernel samples

pling to obtain a sample of files that physically exist in a file cabinet. Suppose 10,000 files are stored in a file cabinet. A sample of 50 files is to be selected for review. Fifty files out of 10,000 files is a rate of one file out of every 200 files. A systematic sampling process starts by selecting a random number between 1 and 200, say 138. Counting through the files, the 138th, 338th, 538th, and so forth, files would be selected for the sample.

In grain inspection, variations of the systematic sampling process are used to select samples. These samples are not random samples by the pure definition, but are approximations from a systematic sampling plan. Risks can be estimated when random samples are taken. If the sampling procedure is not random, or a close approximation, estimates can be biased. One sampling procedure could be to scoop a sample off the top of a lot using a can. If a lot has been loaded and unloaded many times, the lot may be mixed sufficiently that it is fairly uniform and scooping a sample may be adequate. However, some lots may be the combinations of other lots and the resulting lot can be stratified. Scooping a sample off the top may not be very representative of the lot. Commonly accepted sampling procedures have been developed. One detailed source of these sampling procedures is the USDA's Grain Inspection, Packers and Stockyards Administration (GIPSA, 1995a,b). Handbooks provide instructions for sampling from moving grain streams and static grain lots.

The diverter-type (DT) sampler is the most common sampling device for sampling from a grain stream (Figure 2.3). The DT takes a classic systematic sample. The DT traverses a moving grain stream and, per specific timer settings, diverts a small slice of the grain stream to the inspector. The small slices are combined to obtain the sample for the lot.

A manual means of taking a sample from a grain stream, similar to the diverter type sampler, is the pelican sampler. The pelican sampler is a leather bag on the end of a pole (Figure 2.4). A person passes the pelican through a falling grain stream at the end of spout, taking a cut from the grain stream, and emptying the pelican between passes through the grain stream.

The Ellis cup is a manual sampling device for sampling from a conveyor belt (Figure 2.5). A person frequently dips the Ellis cup into the grain stream.

Various probing techniques are used to sample grain from static lots. Depending on the size and shape of the container, multiple probes of the lot are combined to obtain the sample. Patterns for probing a lot are prescribed for various types of containers. The individual

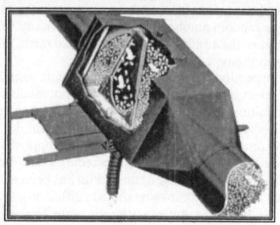

FIGURE 2.3 Schematic of a diverter-type sampler. (*Source:* GIPSA, 1995b.)

FIGURE 2.4. Pelican sampler. (*Source:* GIPSA, 1995b.)

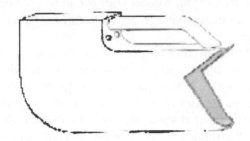

FIGURE 2.5. Ellis cup sampler. (*Source:* GIPSA, 1995b.)

probes are sufficiently close to sample effectively across any stratification that may exist. Figure 2.6 shows an example of a truck probe pattern.

To obtain the specified test sample size, a subsample of the original grain sample must be obtained. Dividers such as the Boerner, Cargo, and Gamet have demonstrated the ability to subdivide an origin sample and have the resulting samples conform to distributions expected from a random process.

The Impact of Sample Size on Risk Management

Measurements associated with grain lots are usually given as percent by weight. The estimates given in the previous discussions expressed the estimates as the percent of the total number of kernels. The percent by kernel count and percent by weight would be the same only if the kernels were all the same size, but kernels are not uniform.

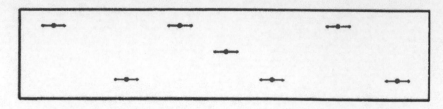

FIGURE 2.6. Example of a truck probe pattern. (*Source:* GIPSA, 1995b.)

Percent by kernel count is, however, usually a reasonable approximation to percent by weight. Kernel counts can be converted to approximate weights by using average kernel weights observed from typical market samples.

The type of measurement is also a consideration in determining the sample size. The analytical methods available for detecting biotech grains may be used to make qualitative or quantitative tests on a sample. A qualitative test can be used to screen lots by providing information on the presence or absence of biotech varieties. Quantitative tests may quantify the total amount of biotech grain, the amount of individual varieties in a sample, or the percentage of biotech or "nonnative" DNA or protein present relative to nonbiotech grain.

When sampling is used in the measurement of some characteristics of a lot, the content of the sample will likely deviate from the lot content. The buyer accepts some risk because the sample may overestimate the quality of the lot. The buyer may assume that the quality of the lot is better than it actually is. Likewise, the seller accepts some risk that the sample may underestimate the quality of the lot. In this case, the seller is delivering better quality than the sample reflects.

Ideally, buyers and sellers would agree to use a sampling plan that provides acceptable risk management. A contract may specify a certain quality level. However, due to sampling variation, a seller may have to provide better quality to have grain lots accepted most of the time. Sellers choose a quality level that they want to have accepted, say 90 percent or 95 percent. This level is sometimes called the acceptable quality level (AQL) (Remund et al., 2001). Likewise, due to sampling variation, the buyer may sometimes have to accept lower quality than the contract specifies. Buyers should choose a lower quality level (LQL) that they want to accept infrequently. This LQL may be acceptable 5 percent or 10 percent of the time. An ideal sample plan would meet both the AQL and the LQL.

An operating characteristic (OC) curve plots the probability of a sampling plan producing a sample that meets an acceptance criterion against the concentration in the lot being sampled. Figure 2.7 gives an example of an OC curve with the ideal relationship of AQL and LQL. A single sample with a qualitative test gives little flexibility for choosing both an AQL and an LQL. Quantitative tests, when available, and multiple sample plans provide more flexibility to choose both an AQL and an LQL.

Sample Size—Qualitative

In reality, all analytical methods have limits of detection. For the purposes of this section, the assumption is that a qualitative test will detect the presence of a single kernel in a sample, regardless of the size of the sample. A positive result does not tell how many biotech kernels are in the sample, only that at least one biotech kernel is pres-

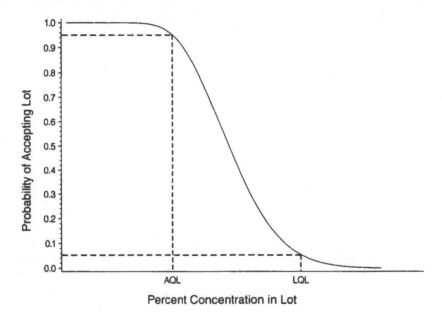

FIGURE 2.7. OC curve with ideal AQL and LQL relationship. (*Source:* Adapted from Remund et al., 2001.)

ent. To choose a sample size, the acceptable and unacceptable concentrations must be decided upon. Since samples are subject to sampling error, acceptable lots may be rejected, and unacceptable lots may be accepted by chance. Buyers and sellers must agree upon acceptable risk.

The following chart shows probabilities associated with different sample sizes (based on kernel count). The horizontal axis gives possible concentrations in lots. The vertical axis gives the probability of observing no biotech kernels in a sample randomly selected from a lot. Curves are given for various size samples.

Figure 2.8 shows probabilities for sample sizes of 60, 120, 200, and 400 kernels. If the desired concentration in the lot is not to exceed 5.0 percent, a sample size of as little as 60 kernels may be satisfactory. Based on a 60-kernel sample, a 95 percent probability exists of rejecting a lot at a 5 percent concentration. If 1.0 percent is the desired maximum concentration in the lot, a sample size of 400 kernels would be more appropriate.

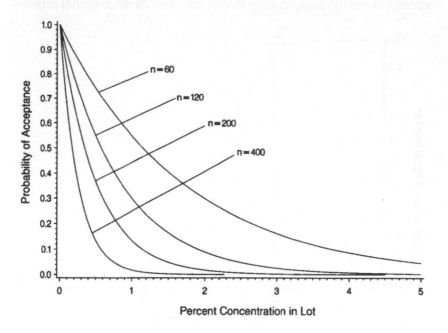

FIGURE 2.8. OC curves for qualitative testing with 60-, 120-, 200-, and 400-kernel samples

Figure 2.9 shows probabilities associated with sample sizes of 800; 1,200; 2,000; 3,000; and 5,000 kernels. Larger sample sizes are used only when low concentrations of biotech kernels are acceptable. For example, if 0.1 percent concentration is the desired maximum, a sample between 3,000 and 5,000 kernels would be recommended. Thus, testing for lower concentrations of biotech grains requires larger sample sizes.

Sample Size—Quantitative

Quantification of the percent biotech grain in a lot is much more problematic than with qualitative testing. As mentioned in the introduction, sampling variability is only one source of error in measurements. Sample preparation and analytical method can be two significant sources of error in the detection of biotech grains. The currently available technologies employed for detection, the polymerase chain reaction (PCR) and enzyme-linked immunosorbent assays (ELISA) have inherent difficulties in producing consistent and accurate quan-

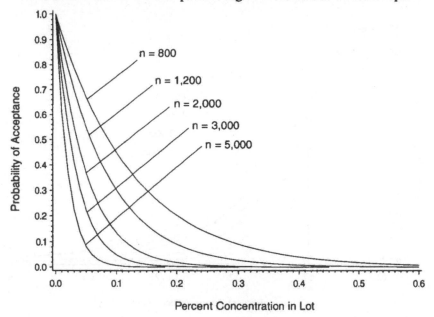

FIGURE 2.9. OC curves for qualitative testing with 800-; 1,200-; 2,000-; 3,000-; and 5,000-kernel samples

titative results. PCR measures the genetic material associated with the inserted DNA. ELISA, on the other hand, measures the protein expressed by the foreign DNA. Both methods present significant challenges in converting the amount of DNA or the amount of expressed protein into the percent of biotech grain by weight. The overall variability of quantitative results, therefore, will be affected by analytical methods as well as sample size. In general, sample size will have little influence on the sample preparation or analytical method. Sample preparation and analytical method are significant sources of error, and increasing the sample size will not reduce the overall variability in measurements as much as expected.

The effect of sample size alone on sampling variability can be examined. The sample is used to estimate the percent concentration in the lot. Using percent by kernel count as an approximation of percent by weight, probability curves can be computed to examine the probability of accepting lots for which a maximum concentration has been specified. Figure 2.10 shows the probabilities associated with a 0.1 percent maximum allowed level of biotech kernels in the sample.

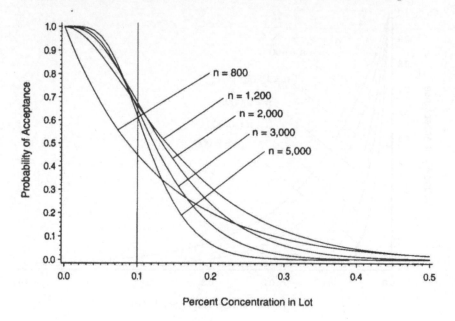

FIGURE 2.10. OC curves for quantitative testing for an acceptance limit of 0.1 percent and sample sizes of 800; 1,200; 3,000; and 5,000 kernels

For this example, a lot with less than 0.1 percent concentration of biotech kernels is defined as acceptable. The curves on the plot give the probabilities of accepting different lot concentrations with five different sample sizes. For concentrations above 0.1 percent, the curves give the probability of accepting an unacceptable lot. This probability may be called "buyer's risk" because this is the chance that the buyer will get an unacceptable lot. For lots with less than 0.1 percent concentrations, the area above the probability curve represents the chance of rejecting an acceptable lot. This may be called the "seller's risk" because it is the chance that the seller will have an acceptable lot rejected.

The ideal sampling plan will minimize both the buyer's and seller's risk. Unfortunately, no one sampling plan will produce both objectives. Increasing the sample size can reduce both buyer's risk and seller's risk. Theoretically, the only limiting factor on the sample size is the lot size. Sample size is often determined by a compromise between the seller's and buyer's risks and the cost of taking and processing a sample. The 5,000-kernel sampling plan in Figure 2.10 allows the buyer to conclude that a 0.2 percent lot has less than a 10 percent probability of being accepted, and the seller to conclude that a 0.05 percent lot has less than a 10 percent probability of being rejected. If the seller is satisfied with an AQL of 0.05 percent, and the buyer is satisfied with a LQL of 0.2 percent, a 5,000-kernel sample with a maximum of 0.1 percent biotech kernels may be an acceptable plan to both parties. Figure 2.11 gives sampling plans that allow a maximum of 1.0 percent biotech kernels in the sample.

Again, increasing the sample size will reduce both the buyer's and seller's risks. Figure 2.12 shows sampling plans that allow a maximum of 5.0 percent biotech kernels in the sample. Estimates for higher percent mixtures will be less precise for the same size sample. However, the buyers and sellers may be willing to accept less precise estimates for these higher mixtures.

Multiple Sample Plans—Qualitative Testing

Previous sections discussed the effects of sample size with qualitative and quantitative testing. One way to express the effect of sample size with qualitative testing is that, for any given concentration, the probability of a negative result decreases as the sample size increases.

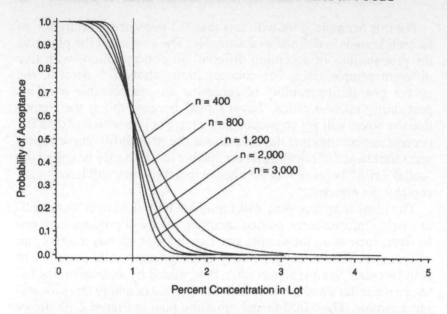

FIGURE 2.11. OC curves for quantitative testing for an acceptance limit of 1.0 percent and sample sizes of 400; 800; 1,200; 2,000; and 3,000 kernels

This is not true only when the lot concentration of biotech grain is zero. If the buyer of a lot has a zero tolerance for biotech grain, taking the largest sample possible will best serve the needs of the buyer. The buyer can be reasonably certain that a lot with a high concentration of biotech grain will be rejected.

A single large sample serves the buyer's interests well. However, some buyers may be willing to accept some low concentrations but unwilling to accept high concentrations. Sellers of lots with low concentrations would like to have high probabilities of testing negative. Decreasing the sample size will increase the chances of a negative result on low concentrations. Unfortunately, decreasing the sample size increases the chance of a negative result with higher concentrations. When a low concentration is acceptable to the buyer, a single qualitative test may not serve the interests of both the buyer and the seller. An alternative is to implement a multiple sample plan.

Multiple sample plans specify that a certain number of independent samples will be selected from the lot and tested. The buyer will

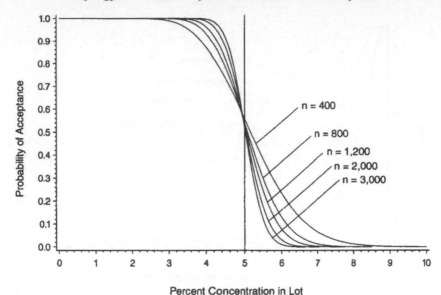

FIGURE 2.12. OC curves for quantitative testing for an acceptance limit of 5.0 percent and sample sizes of 400; 800; 1,200; 2,000; and 3,000 kernels

accept the lot if certain combinations of positive and negative test results are obtained. For example, the sample plan may specify that five samples of 100 kernels will be selected from a lot. If no more than three positives are obtained on the five tests, then the lot is acceptable.

The components of a multiple sample plan are the number of samples, the size of each sample, and the maximum number of samples testing positive. Changing any one or more of these parameters affects the probability of acceptance. Buyers and sellers can choose a plan based on the risks they are willing to assume, as well as the cost of conducting the tests.

With the example of five samples of 100 kernels, the maximum number of positives specified in the plan could be one, two, three, or four. Figure 2.13 gives the probabilities of accepting lots with these four plans. Increasing the maximum number of acceptable positives will result in higher concentrations being accepted. By manipulating all three parameters, the shape of the probability curves can be changed considerably. Figure 2.14 (Remund et al., 2001) compares a

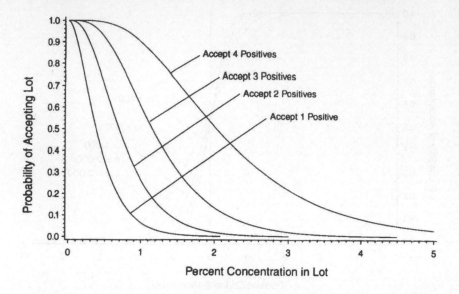

FIGURE 2.13. OC curves for a multiple sampling plan that tests five samples of 100 kernels and accepts from one to four positives

six-sample plan with a 60-sample plan. Both plans have low probabilities of accepting a 1.0 percent concentration. The probability of accepting a 0.5 percent concentration sample, however, is considerably higher for the 60-sample plan than for the six-sample plan. Multiple sampling plans can be used to balance the risks between buyers and sellers when low concentrations are acceptable to buyers.

Sample Preparation: Obtaining a Portion for Analysis

The discussions on sample size have been presented thus far as if each kernel in the sample was measured independently. The major analytical technologies for detecting biotech grains, PCR, and ELISA usually do not process individual kernels, but rather make a measurement on a preparation from the sample. An 800-kernel sample of corn will weigh more than 200 grams. Some technologies cannot make measurements on the entire sample. Sometimes, one gram or less can be measured at a time. In a typical process, the entire sample of grain

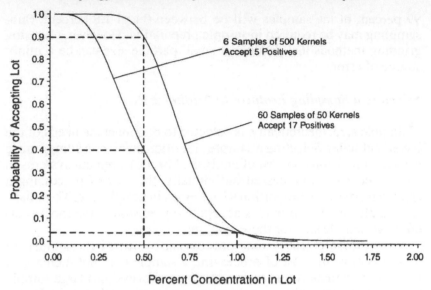

FIGURE 2.14. OC curves comparing two multiple sampling plans. (*Source:* Adapted from Remund et al., 2001.)

will be ground and a subsample taken for further processing and measurement.

As mentioned in the introduction, sampling variability is only one source of error in measurements. Sample preparation and analytical methods are two other significant sources of error. Grinding and subsampling are parts of the sample preparation and contribute to preparation error. Subsampling particles from the ground sample can be modeled as a statistical sampling problem, just as sampling kernels from the lot.

Subsampling error can be minimized with a grinding procedure that produces a small particle size. If corn is ground to spherical particles of 200 microns in diameter, the calculated number of particles is more than 180,000 per gram, assuming a density of 1.3 grams/cm^3 (U.S. Grains Council, 2002). When subsampling from a sample with 1 percent biotech material, 99 percent of the subsamples would be between 0.94 and 1.06 percent biotech. Relative to the error associated with sampling from the lot, the subsampling error is relatively minor. For sampling 800 kernels from a lot with 1 percent biotech kernels,

99 percent of the samples will be between 0 and 1.8 percent. Sub-sampling may be required in sample preparation. However, choosing grinding methods that produce a small particle size can be a minor source of error.

Selecting a Sampling Protocol to Minimize Risk

Sample size, theoretically, is selected to best meet the needs of the buyer and seller. Selecting a sample size often involves a compromise between precision and cost of analysis. For measurement systems in which kernels are processed individually, the cost of processing a sample increases in proportion to increases in sample size. The smallest sample size that provides acceptable precision is the most cost-effective sample size for these systems.

Many measurement systems process and measure bulk samples. In these systems, the cost of processing a sample may not increase in proportion to increases in the sample size. Processing a large sample may cost only slightly more than processing a small sample. Under these circumstances, processing the largest sample the technology can accommodate may be the best sample size.

For single-sample qualitative testing, sample sizes can be determined with a relatively simple formula. Given the desired lot concentration and probability of detection, a sample size is computed with the following formula:

$$n = \log \left[1-(G/100)\right]/\log \left[1-(P/100)\right]$$

where n is the sample size (number of kernels), G is the probability (in percent) of rejecting a lot concentration, and P is the percent concentration in the lot.

Tables 2.1 and 2.2 provide recommended sample sizes for qualitative testing based upon this formula. Therefore, as can be seen from Table 2.1, a representative sample of 299 kernels/beans from a lot containing 1.0 percent biotech grain will contain one or more biotech kernels/beans 95 percent of the time.

For very low values of lot concentration, the sample size may become very large. Suppose someone wants to detect a 0.01 percent lot concentration with a 99 percent probability. The required sample size would then be 46,050 kernels. Such a large sample, however, may not be appropriate for use with all testing methods. The available testing

TABLE 2.1. Sample sizes such that lots containing the given concentration levels are rejected 95 percent of the time

Biotech Concentration	Number of Kernels	Approximate Weight in Grams	
		Corn	Soybeans
0.1	2995	881	474
0.5	598	176	95
1.0	299	88	48
2.0	149	44	24
3.0	99	30	16
4.0	74	22	12
5.0	59	18	10

TABLE 2.2. Sample sizes such that lots containing the given concentration levels are rejected 99 percent of the time

Biotech Concentration	Number of Kernels	Approximate Weight in Grams	
		Corn	Soybeans
0.1	4603	1354	729
0.5	919	271	146
1.0	459	135	73
2.0	228	68	37
3.0	152	45	25
4.0	113	34	18
5.0	90	27	15

technology may not be able to process such a large sample, or may not have the sensitivity to detect one biotech kernel in 46,050. A sample that is too large may be tested by dividing it into equal subsamples of smaller size. Each subsample will require testing, and none of the subsamples may have a positive result. For example, a test kit may be able to detect one biotech kernel in 1,000 kernels with a high degree of reliability, but not have the sensitivity to detect a 0.01 percent concentration. To achieve an appropriate sample size for the test kit, the 46,050-kernel sample can be divided into 47 subsamples of almost 1000 kernels each. All 47 subsamples would require testing. For the lot to be acceptable, all subsamples must be negative.

CONCLUSIONS

When a sample is used to represent the content of a lot, the content of the sample will likely deviate from the actual content of the lot. Sampling error creates risks for both the buyer and seller in a transaction. The buyer may get a lot with a higher concentration than desired. Likewise, the seller may have a lot rejected that actually meets contract. Probability can be used to estimate a likely range that a sample may deviate from the true lot content, as long as the sample is properly taken. Knowledge of these probabilities allows both the buyer and seller to manage marketing risks by choosing appropriate sampling and testing strategies.

REFERENCES

Ahmed, F.E. (1999). Safety standards for environmental contaminants in foods. In *Environmental Contaminants in Foods* (pp. 55-570), C. Moffat and K.J. Whittle, eds. Sheffield Academic Press, United Kingdom.

Animal and Plant Health Inspection Service (APHIS) (2002). *United States Regulatory Oversight in Biotechnology* <http//www.aphis.usda.gov/biotech/OECD/usregs.htm>.

U.S. Food and Drug Administration (FDA) (1992). Foods derived from new plant varieties. *Federal Register* 57:22984-23005 (May 29).

General Accounting Office (GAO) (2002). Experts view regimen of safety tests as adequate, but FDA's evaluation process could be enhanced. Report to congressional requesters. U.S. General Accounting Office, Washington, DC, May.

Grain Inspection, Packers and Stockyards Administration (GIPSA) (1995a). *Handbook—Book 1*. Washington, DC. <http://www.usda.gov/gipsa>.

Grain Inspection, Packers and Stockyards Administration (GIPSA) (1995b). *Mechanical Sampling Systems Handbook*. Washington, DC. <http://www.usda.gov/gipsa>.

Hogg, R. and Craig, A. (1978). *Introduction to Mathematical Statistics*. Macmillan Publishing Co., New York.

McLeod, A.I. (1988). Simple random sampling. In *Encyclopedia of Statistical Science* (pp. 478-480), Volume 8, S. Kotz and N. Johnson, eds. John Wiley & Sons, New York.

Remund, K.M., Dixon. D.A., Wright, D.L., and Holden, L.R. (2001). Statistical considerations in seed purity testing for transgenic traits. *Seed Sci. Res.* 11:101-119.

U.S. Grains Council (2002). *2001/2002 Value Enhanced Grains Quality Report.* <http://www.vegrains.org/documents/2002veg_report/veg_report.htm>.

Whitaker, T.B., Freese, L.F., Giesbrecht, F.G., and Slate, A.B. (2001). Sampling grain shipments to detect genetically modified seed. *J. AOAC Int.* 84(6):1941-1946.

Chapter 3

Sampling for GMO Analysis: The European Perspective

Claudia Paoletti

INTRODUCTION

Scientific Background

The use of genetically modified organisms (GMOs) in food products has increased steadily since the first experiments on transgenic organisms were carried out during the 1970s on bacteria, and on both plants (first tobacco plant modified with *Agrobacterium*) and animals (first mouse inserted with genes regulating human growth hormones) in the 1980s.

Since the 1980s the possibility of introducing novel traits into crop cultivars allowed plant breeders to implement numerous advances beyond any feasible expectation. Today transgenic plants are a reality that provides new opportunities to increase yield and overall food production at a rate comparable to the escalating rise of global food-market demand. Yet the use of transgenic plants has always been linked to extensive controversy regarding the large-scale commercial release of GMO crops, both in the public (Bowes, 1997; Ellstrand, 1993; Krawczyk, 1999; Philipon, 1999) and scientific (Colwell et al., 1985; Dale, 1994; Dale et al., 2002; Ellstrand et al., 1999; Van Raamsdonk, 1995) domains. These concerns can be traced back to

I am very grateful to Michael Kruse, Simon Kay, Peter Lischer, and Dorothée André for all our interesting discussions on sampling. Thanks to Heidi Hoffman, Paolo Pizziol, and Guy Van den Eede for reviewing a previous version of this manuscript. I am particularly grateful to Marcello Donatelli for his scientific support and valuable intellectual contribution.

two basic fundamental questions: Are GMOs and GMO products dangerous for human health? Are they dangerous for the environment?

The ongoing debate regarding the safety of genetically modified (GM) food products consumption centers on a few important arguments, extensively discussed in the pertinent literature (Belton, 2001; Kuiper et al., 2001; FAO/WHO, 2001), which can be summarized as the following:

1. Possible toxicity according to the nature and function of the newly expressed protein
2. Possible unintended secondary effects linked to the insertion of new genes due to the limited knowledge of gene regulation and gene-gene interactions (pleiotropic effects) in plant and animal genomes
3. Potential for gene transfer from GM foods to human/animal gut flora
4. Potential allergenicity of the newly inserted trait

The definition of internationally harmonized strategies for the evaluation of GMO food safety has been a priority since biotechnological food production systems started in the early 1990s. Several organizations—the International Food Biotechnology Council (IFBC), Organization for Economic Cooperation and Development (OECD), Food and Agriculture Organization of the United Nations (FAO), World Health Organization (WHO), and the International Life Science Institute (ILSI)—have developed guidelines for novel food safety assessment (reviewed in Kuiper et al., 2001). Most of these guidelines are based on the "concept of substantial equivalence," which states that existing traditional foods can serve as counterparts for comparing the nutritional properties and safety level of novel GM foods (FAO/WHO, 1991). However, despite the field's progress, the application of the concept of substantial equivalence needs further elaboration and international harmonization with respect to critical parameters, requirements for field trials, data analysis approaches, and data interpretation in order to provide scientifically reliable results (Kuiper et al., 2001). More in general, the identification of the long-term effects of *any* food, traditional or transgenic, is very difficult due to the many confounding factors and the great genetic variability in food-related effects among different populations. In this context it is not surprising that the lack of historical record on the potential effects of GM foods consumption is still sustaining public concerns linked to the use of GM products in food production systems. This lack of documentation on GM food safety becomes even more evident when compared to the long-term experience and history of safe use of foods resulting from traditional crop breeding and selection programs.

From an environmental perspective, the concerns associated with the commercialization of transgenic plants are due to the risk of gene flow from crops to wild or weedy relatives (Dale et al., 2002; Darmancy et al., 1998; Darmancy, 1994; Ellstrand and Elam, 1993; Simmonds, 1993; Snow, 2002). Even if crops and weeds have exchanged genes for centuries, the coexistence of transgenic plants with nontransgenic wild and/or cultivated species raises several issues that are extensively described and discussed in the pertinent literature.

The assessment of the environmental impact of GM crops is a fundamental part of the international regulatory process undertaken before GM crops can be grown in field conditions. Nevertheless, a review of the current literature on the possible long-term consequences of GM-crops-to-weed hybridization (Dale et al., 2002) indicates that our understanding of these topics is still very limited and rudimentary. In opposition to these uncertainties, well-established scientific evidences indicate that preventing gene flow between sexually compatible species in the same area is virtually impossible (Dale, 1994; Ellstrand et al., 1999; Ellstrand and Hoffman, 1990; Slatkin, 1985; Slatkin, 1987). As a result, the real challenge facing current genetic research is deciding what constitutes an acceptable or unacceptable environmental impact.

This chapter focuses on one critical aspect of GMO control in market products: the definition of sampling protocols for GMO detection and/or quantification. Current legislation in the European Union (EU) requires the labeling of food products, raw or processed, that contain GM material. Despite being focused on the demands of the European consumer, this requirement has wider consequences affecting all major food producers worldwide. As a result, large-scale testing and monitoring programs are being conceived and executed in order to check for compliance with the regulations involved. Therefore, there is a strong interest in the schemes adopted for the sampling of food products to ensure accuracy and precision of GM testing surveys.

EU GMO-Related Legislation

The legislation regulating GMOs and GMO-derived products in the European Union is formally structured into a horizontal (background legislative framework) and a vertical (sector-specific) legislation, designed to provide a new set of legislative tools according to the specific needs and requirements of each sector involved. Two directives define the horizontal legislative framework: Directive 2001/18/EC (concerning the deliberate release of GM organisms) entered into force on October 17, 2002, replacing Directive 90/220/EEC, and Directive 98/81/EC (concerning the contained use of GM microorganisms), which amended Directive 90/219/EEC.

Directive 2001/18/EC puts in place a step-by-step process to assess possible environmental and health risks on a case-by-case basis before granting the authorization for experimental releases (part B of the directive) and marketing (part C of the directive) of GMOs or their derived products. The new Directive 2001/18/EC upgrades Directive 90/220/EEC mainly by expanding public information opportunities, by introducing general rules for traceability and labeling, and by strengthening the decision-making process to assess environmental risks linked to the release of GMOs.

Both the old and the new directives (art. 16 in Directive 90/220/EEC, art. 23 in Directive 2001/18/EC) allow member states to ban the introduction and/or cultivation of authorized GMOs if new or additional information identifies specific risks to human health or the environment. Article 16 has been invoked in several instances. After detailed examination of the dossiers by the relevant European scientific committee, the European Commission rejected a large number of such cases. However, a de facto moratorium exists on the entire authorization process. Indeed, only the enforcement of Directive 2001/18/EC is going to boost the lifting of the moratorium and restart the overall authorization process.

Vertical legislation has been issued for novel foods and novel foods ingredients (Regulation EC No. 258/97). Under Regulation EC No. 258/97, the authorization for GMO products must also be based on a safety assessment. An alternative simplified procedure may be applied for foods derived from GMOs, but no longer containing GMOs, if "substantial equivalence" to existing foods can be demonstrated. So far the marketing of products derived from GMOs, such as oils, flours, and starches was based on this simplified notification procedure, claiming for the substantial equivalence of the products to their "conventional" homologues.

The European Commission (EC) has proposed to replace the authorization procedure for GMO foods as laid down in the novel foods regulation with a streamlined and more transparent community procedure for all marketing applications, whether they concern the "living" GMO as food and feed or GMO-derived food and feed products. In the new regulation, the simplified alternative will be abandoned and only one standard procedure will be allowed.

In addition, the new regulation will recast the provisions concerning GM food labeling, so far dealt with in the novel foods regulation and for two specific products in Council Regulation 1139/98, amended in January 2000 by the so-called threshold regulation, Commission Regulation 49/2000. According to the current rules, the labeling of food and feed genetically modified or produced from GMOs is mandatory if the genetic modification can be detected in the final product, unless the presence of GM material is below a threshold of 1 percent. However, this threshold exists only for adventitious and technically unavoidable traces. In the debate on the new commission proposals, the European Parliament is considering lowering this threshold, but a final decision has not been made. Yet the setting of a threshold value is of utmost importance in the overall GMO debate because all the steps of GMO control programs (definition of sampling protocols, methods development, and methods validation) are strongly affected by the enforced threshold.

Vertical legislation for seeds and propagating material is also in place (Directive 98/95/EC) and a discussion is going on to reach a consensus on the thresholds for the adventitious and technically unavoidable presence of GM kernels in non-GM kernel lots. The actual EC seed marketing legislation, focusing on certification requirements and procedures, remits to Directive 2001/18/EC for the environmental risk assessment of GM seeds and grains.

THE CONCEPT OF SAMPLING AND CURRENTLY
ADOPTED SAMPLING PROTOCOLS

Sampling

A sample is a collection of individual observations selected from a population according to a specific sampling procedure. A population is the totality of individual observations about which inferences are to be made. By definition, the process of sampling is always a source of error (sampling error) when estimating population characteristics. Ideally, all the observations of a population should be included in a survey in order to ensure maximum accuracy and precision of the results (estimate = real value). Such an approach is practically inconceivable, and sampling theory was developed precisely to define suitable strategies for obtaining reliable estimates from limited numbers of measurements. Therefore, the goal of "good" sampling practice is to minimize the unavoidable sampling error, by ensuring that the sample is representative of the entire population. As a result, the first critical step in the planning and execution of any survey is a precise definition of the population of interest in order to choose the optimal sampling strategy as a function of population properties.

Although these intuitively simple concepts constitute the basis of sampling theory and are extensively addressed in the pertinent literature (e.g., Cochran, 1977), they represent a real challenge for most surveys. Defining the population to be sampled can be easy (e.g., take a sample from 5,000 cards), or very difficult (e.g., take a sample of the chocolate bars sold in European supermarkets) depending upon the amount of information initially available for fixing population boundaries (i.e., population size and spatial structure) and assessing its characteristics (e.g., the distribution of the variable of interest).

In the case of GMO detection and/or quantification in different market products, population definition can be very tricky. In general, a GMO analytical service is carried out to gain information regarding the composition of a large body of target material (e.g., lots of kernels, ingredients, or final food products). In contrast, a very small amount of material is subject to the analytical procedure. As a consequence, the sampling processes used to prepare the analytical sample from the target material are key to reliable and informative analysis of foods and agricultural products. Indeed, the effort of the analyst in the laboratory is futile if sampling has not been carried out correctly. Analytical reliability is today limited not by the analyst intrinsic qualities, but by the lack of reliability of the samples submitted to the analytical process. Quality estimation is a chain and sampling is by far, its weakest link. (Lischer, 2001a)

Within Europe, large-scale testing and monitoring programs are being conceived and executed in order to check for compliance with EU legislation requirements for labeling of food products containing GM material. Several guidelines defining sampling strategies for quality and purity analyses are already available from standards authorities and national and international organizations. Some of these sampling protocols have been provisionally adopted for the detection of GM materials while control authorities are waiting for the ad hoc protocols for GMO detection that are currently being conceived by several expert groups.

Sampling Protocols for Raw Bulk Materials

Several national and international organizations have developed and recommended approaches for kernel (i.e., seeds and grains) sampling, including

1. International Seed Testing Association (ISTA): *Handbook on Seed Sampling* and the *International Rules for Seed Testing;*
2. U.S. Department of Agriculture, Grain Inspection, Packers and Stockyards Administration (USDA GIPSA): *Grain Inspection Handbook;*
3. Comité Européen de Normalisation (CEN): draft standard on sampling for GMO detection;
4. European Union: Aflatoxin directive 98/53;
5. WHO-FAO *Codex Alimentarius:* proposed draft general outline on sampling;
6. International Organization for Standardization (ISO): standards 13690 and 542.

Despite many similarities, a systematic comparison of these sampling procedures has revealed several critical differences (Table 3.1) among the approaches chosen by the various organizations (reviewed

TABLE 3.1. Comparison of the main sampling protocols adopted for GMO analyses

Source	Lot size	Increment	Increment size	Bulk	Sample reduction	Sample preparation	Laboratory
ISTA IRST	According to species: 10,000 kg to 40,000 kg (max)	One increment per 300 kg to 700 kg	Not indicated	1 kg	Repeated halving or random combinations	Not generally applicable to seed quality	1 kg (for analysis of contamination by other seed varieties)
USDA GIPSA	Up to 10,000 bushels (~254,000 kg), or 10,000 sacks if the lot is not loose	3 cups or 1 cup per 500 bushels (~12,000 kg)	1.25 kg	Equivalent to laboratory sample	"Suitable" dividing tools	Whole laboratory sample in food blender	~2.5 kg, but not less than 2 kg
ISO 13690, static sampling	Up to 500,000 kg	15 to 33 for loose bulk (up to 500,000 kg)	Not specified	Not indicated	Examples of dividing tools given	Not covered by this document	>1 kg (for kernels)
ISO 542, oil seed sampling	Up to 500,000 kg	15 to 33 (static, 50,000 kg), "as many as possible" free-flowing	Stated as 0.2 kg to 0.5 kg, but in practice up to 5 kg	100 kg	Examples of dividing tools given	Not covered by this document (cf., ISO 6644)	2.5 kg to 5 kg
EU Dir. 98/53	No limit if not separable, otherwise up to 500,000 kg	Up to 100 kg	0.3 kg	30 kg if lot size ≥ 50,000 kg, 1-10 kg if lot size <50,000 kg	Mixed and divided (no examples)	Should be homogenized "with care"	10 kg
CEN	Up to 500,000 kg	As for ISO 13690, except the number of depths per sample point not specified	Started as 0.5 kg, but in practice up to 5 kg, if ISO 542 fully applied	20 times laboratory sample (i.e., 60 kg)	Follows ISO 13690	Not covered by this document	10,000 kernels
WHO-FAO	Discussed, but not specified	Not indicated	Not indicated	Various populations	Not discussed	"Appropriate" crushing, grinding, and dividing procedures	Not discussed

Source: After Kay, 2002.

by Kay, 2002). However, in order to critically evaluate the protocols it is necessary to briefly describe, in general terms, the practical implications of kernel lot sampling.

Kernel lot sampling is a multistage procedure that reduces a lot to a much smaller amount of material, the analytical sample, which is analyzed in the laboratory (Kay and Paoletti, 2002). In Figure 3.1 a general scheme of the sampling steps involved in the creation of the analytical sample is shown: a given number of increments (see Table 3.2 for technical definitions) are sampled from the lot, according to a specific sampling plan, and combined to produce a bulk sample. A fraction of the bulk sample, the laboratory sample, is randomly sampled and sent to the laboratory, where it is ground to a fine powder. A portion of the laboratory sample, the test sample, undergoes the analytical process after being randomly split into portions (analytical samples) that can be wholly analyzed in a test reaction. The objective of this long and complex sampling process is clear: produce a final sample of suitable working size that truly represents the lot. As evident from Table 3.1, the recommended sampling strategies differ substan-

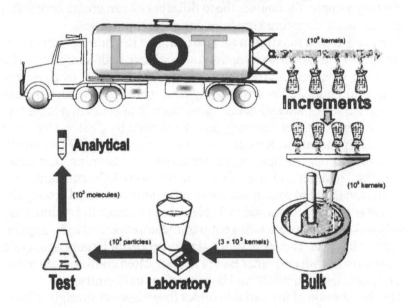

FIGURE 3.1. Schematic representation of all the sampling steps involved in the creation of the analytical sample. As an example, sample sizes for maize are indicated in parentheses.

TABLE 3.2. Definition of technical terms used for kernel lot sampling, as applied in the text

Term	Definition
Increment sample	A group of kernels collected simultaneously from one location in the lot
Bulk sample	The combination of all increments sampled from a given lot
Laboratory sample	The fraction of the bulk sample processed by the laboratory for analysis
Test sample	The fraction obtained from the splitting of the laboratory sample into portions for the purposes of the analysis
Analytical sample	The portion of the test sample that is wholly analyzed in the test reaction

Source: Adapted from Kay (2002)

tially in terms of maximum allowed lot size, number of increments used to create the bulk sample and number of kernels present in the laboratory sample. Of course, these differences can create confusion when choosing sampling strategies for control purposes.

The need to fix a maximum lot size is linked to a number of considerations and statistical assumptions associated with sampling. The application of specific sampling plans to lots larger than those expected would imply a lower sampling rate, which may reduce the expected accuracy of the estimates. In addition, it appears that large lots are more likely to show heterogeneity (reviewed by Coster, 1993; but see also Jorgensen and Kristensen, 1990; Tattersfield, 1977), which could compromise sample representativeness, assuming that other factors (i.e., number and size of increments) were held constant. Fortunately, the range of maximum lot sizes is quite restricted among the different protocols compared in Table 3.1. The lower ISTA limits are clearly linked to the more stringent parameters expected to be applied to seeds. However, the issue of lot size is probably largely academic by comparison with two other issues that are often confused: distribution of particles within lots and degree of lot uniformity.

The distribution of the variable under investigation strongly affects sampling strategy requirements. The vast majority of kernel sampling plans, including those compared in Table 3.1, are based upon the assumption of random distribution of GM material so that the mean, the standard deviation of the mean, and both the producer and consumer risks can be estimated according to the binomial, the Poisson, or the hypergeometric distributions (ISO, 1995; see also Remund et al., 2001, for GMO discussion; and Cochran, 1977, for statistical discussion). Nonetheless, assuming randomness is very risky (Paoletti et al., 2003; Lischer, 2001a; Gilbert, 1999; Kruse, 1997). As explained by Lischer (2001a), many sampling mistakes are linked to the confusion between populations where neighboring particles are totally independent from one another (random distribution) and populations where neighboring particles have a given probability of being correlated (heterogeneous distribution). Experience shows that perfect disorder is the exception and partial order is the rule. In the case of raw bulk materials, such as kernels, we must take into account that industrial activities are well framed in time and space. This generates correlations that, ultimately, will promote segregation during transportation and handling of the material.

The other important factor affecting sampling requirements is the degree of lot uniformity. A lot is uniform if its composition is spatially constant. In theory, a truly uniform lot would require only a single increment sample to estimate the proportion of GM material. However, in practice, a considerable amount of particulate material is rarely uniform: the more nonuniform the material is, the more difficult the sampling operation necessary to ensure sample representativeness. All protocols recognize that lot uniformity cannot be assumed on an a priori basis and recommend to include more than one increment per lot (see Table 3.1).

Hence, the real challenge in the definition of "good" kernel sampling protocols is establishing *how* and *how many* increments must be sampled to ensure representativeness. *How:* without exception, all protocols recognize that sampling from flowing material during lot loading or unloading is the best choice. Indeed, sampling increments from a flowing stream is an effective strategy to maximize sampling efficiency in the case of heterogeneity (Cochran, 1977; see Lischer, 2001a, for specific GMO discussion). However, it is also recognized that this is not always possible, and much discussion of procedures for sampling static lots is included in the protocols. *How many:* the number of increments depends upon the expected distribution of GMO in the lot. As heterogeneity increases, the number of sampled increments should also increase. The wide range of the recommended increment numbers (see Table 3.1) illustrates the different levels of pragmatism accepted by the various committees involved. Unfortunately, although acknowledging that the recommended procedures are not effective in the sampling of nonrandom distributions, the documents compared in Table 3.1 do not specifically address these issues.

Another substantial difference among kernel sampling protocols is linked to the recommended size of the laboratory sample (see Table 3.1). Ideally, the laboratory sample should correspond perfectly in terms of proportion of GM/non-GM material to that of the original bulk sample. This is nominally achieved through a simple random sampling process, usually by appropriate mixing followed by division of the sample into increasingly smaller portions (Kay and Paoletti, 2002). Although most guidelines discuss the use of dividers or other approaches for splitting samples, discrepancies exist among protocols regarding the final laboratory sample size. Indeed these discrepancies are one of the main focuses in the ongoing debate on sampling.

In summary, the action of sampling from a kernel lot to produce an analytical sample is a multistep process that requires careful respect of the various statistical assumptions implicit in each of the steps involved.

Sampling Protocols for Primary Ingredients and Complex Food Products

The definition of suitable sampling plans for GMO detection/quantification in primary ingredients and complex food products is much more complex compared to raw bulk materials because of the tremendous difficulties encountered in defining population properties and, ultimately, ensuring sample representativeness. On occasion, food product lots are assumed to be uniform in order to overcome the problem of sampling plans definition and to extrapolate broader statements from the results of local analyses. However, such an approach is dangerous and statistically objectionable (Kay and Paoletti, 2002; Paoletti et al., 2003). Some existing ISO standards (e.g., ISO 2859, 3951, 8422, 8423, 8550, 10725) deal specifically with issues related to the sampling of food products, agricultural commodities, and bulk ingredients. The idea of adopting these sampling recommendations for GMO detection/quantification has been put forward. However, it must be recognized that such protocols were not intended for GM materials analyses: the diverse nature of the target analyte, the concentration and distribution in the product, the production and packing processes, the distribution channels are just a partial list of all the factors that preclude the possibility of standardizing sampling plans for primary ingredients and, particularly, final food products. Without a doubt, the problem is complex and has no easy solution.

Two specific perspectives can be taken with respect to ingredients and complex food products sampling. The first is what Lischer (2001a) defines as "grab sampling": final food products are "randomly" selected without consideration of the population from which they are drawn (e.g., packets of biscuits taken from supermarket shelves). Such an approach has been adopted occasionally on the basis that it could be regarded as simple random sampling. Clearly, this is not the case with "grab sampling" because the elements of the population do not have an equal probability of being selected during the sampling process, a prerequisite to the correct application of simple random sampling (Cochran, 1977). As a consequence, with such approach any statement concerning GM content can only apply to the selected packet: no statement can be made about all the other packets that have not been submitted for testing. Indeed, this constitutes an excellent example of how a selection process can lack any sampling framework.

Alternatively, efforts (in terms of both time and financial costs) could be made to ensure that the analysis of food market products populations is based on correct sampling plans (Kay and Paoletti, 2002). Unfortunately, a generalized sampling protocol that fits most circumstances cannot be defined. As a matter of fact, an effective coordination between those who are involved in the production chain and statisticians is mandatory to ensure that the collected sample is representative. For example we may choose to test the following:

- Chocolate-coated cookies (all types)
- Produced by manufacturer Y in its factory at Z
- Between the dates of September 1, 2002, and September 15, 2002
- On sale in region X

This statement will then allow the statistician to define the appropriate sampling framework, specifically *where* and *how many* packets of cookies must be selected. However, these pieces of information alone are not sufficient to define population properties, and more feedback from those involved in the preparation and commercialization of the product is necessary:

- Where are such cookies sold?
- What is the list of chocolate-coated cookies produced by manufacturer Y?
- What is the expected turnover of different distributors? Should higher turnover distribution points be sampled more intensively?
- What is the total number of packets distributed into region X that were produced in the time period concerned?

Once this information is available, the statistician might consider it necessary to define a number of sampling *strata*. Broadly speaking, if a distribution outlet—e.g., a big supermarket chain with a given number of stores—has a high proportion of the turnover, then more samples should be drawn from those few stores. Conversely, fewer samples would be collected from smaller outlets that sell fewer cookies. All these difficulties linked to the definition of population properties are the reason why the definition of sampling plans for GMO detection in final food products is still an issue that needs to be solved, despite the many efforts of standard committees and expert groups worldwide.

LIMITATIONS OF THE CURRENTLY ADOPTED SAMPLING PROTOCOLS

Most of the following discussion will focus on the sampling protocols currently adopted by the EU member states for kernel lot analysis. Indeed, as explained previously, no official sampling protocols for final foods analysis exist that can be compared and discussed. Most kernel lot sampling protocols (Table 3.1) rely upon three assumptions:

1. From a statistical point of view, the sampling procedure is considered to be a single operational step, even though it may comprise a series of individual, independent actions.
2. No consideration is made for the statistical implications of the typical clustered sampling of kernels (increments) applied in practice to the lot.
3. The binomial (or Poisson) distribution is used to calculate the precision of the estimate or, in other words, the consumer and the producer's risk.

Assumption One

As illustrated in Figure 3.1, kernel lot sampling implies a multistep procedure that reduces the lot to an analytical sample. Each of the sampling steps involved is, by definition, a source of sampling error that will contribute to the overall sampling error associated to the sampling process. In addition to the unavoidable sampling error (SE), each sampling step has an associated preparation error (PE) determined by the handling routine to which the sample is submitted during the analysis (mixing, drying, filtering, grinding, etc.). In the case of GMO analysis (see Figure 3.1) the laboratory sample (kernels) is ground to particles and DNA molecules are extracted from the test sample. According to Pier Gy's sampling theory (Lischer 2001a), the total sampling error (TSE) associated to each step can be calculated as

$$TSE = SE + PE.$$

If there are N sampling and preparation stages, N total sampling errors will be generated, and if AE is the analytical error intrinsic to the analytical method used for the laboratory analysis, the overall final error is defined as

$$OE = AE = \sum_{n=1}^{N} \left(SE_n + PE_n \right).$$

Clearly, if the sampling procedure is statistically treated as a single operational step, the TSE is going to be strongly underestimated.

Assumption Two

Since, for practical reasons, the selection of kernels cannot be performed in an independent manner (i.e., simple random sampling of individual kernels), a bulk sample is first produced by systematically sampling scoops of kernels, the increments. The clustered nature of this first sampling step must be taken into account when estimating the overall error associated with any given protocol (Paoletti et al., 2003). Statistically speaking, the sampling error associated with the sampling step of lot to bulk would be the minimum possible only if the grains making the bulk sample were collected one by one. As soon as increments are sampled, these conditions are violated and an additional error is introduced: the grouping error. The larger the increment size, the larger the grouping error becomes.

Of course the extent to which the grouping error compromises the accuracy of the estimate depends also on the degree of lot heterogeneity. As discussed previously, we must distinguish between populations in which neighboring particles are totally independent from one another (random distribution) and populations in which neighboring particles have a given probability of being correlated (heterogeneous distribution). The more heterogeneous the distribution is in the lot, the less representative is the bulk sample created from the mixing of a few clusters of grain.

Assumption Three

Kernels, including GM contaminations, are assumed to be randomly distributed within lots so that the binomial distribution can be used to estimate both the producer and consumer's risk. However, the assumption of random distribution is likely to be false. A population (lot) of particulate material (kernels) is always affected by a certain amount of heterogeneity. According to Lischer (2001a), the degree of heterogeneity of a lot can be described as a scalar function, for which the state of homogeneity is a limit case. Experimental confirmation of this theory comes from several studies investigating the degree of heterogeneity for several traits in large seed lots (Tattersfield, 1977; Jorgensen and Kristensen, 1990; Kruse and Steiner, 1995). Extensive heterogeneity has also been reported for kernel lots produced with large-scale facilities, such as those for grass seed production in the Midwest. A disturbing explanation offered by some authors is that "such seed lots are seldom if ever blended by state-of-the-art equipment, but are simply conditioned, bagged, and marketed" (Copeland et al., 1999).

Attempts to adapt the mathematical properties of Poisson distribution to events of nonrandom GM material distributions have been made (Hübner et al., 2001). However, such approaches may violate inherent assumptions (e.g., normal variance characteristics) required for the use of such tools. Lischer (2001a,b) has attempted to address these issues treating the sampling process as a series of error components that are estimated empirically, taking into account grouping of increments for various levels of heterogeneity. Nevertheless, the approach requires strong assumptions on the expected variability of the various parameters, which are difficult to confirm without experimental data.

An additional difficulty is given by the expected low percentage of impure kernels per lot. The estimation of such low proportions is a classical problem in sampling theory, usually addressed by the assumption concerning the nature of the distribution of the material (Cochran, 1977). As noted earlier, however, knowledge of the true likely distribution of GM material in kernel lots is extremely impracticable and costly to obtain.

So far I highlighted that the first sampling step (lot to bulk) is the most problematic one among all those applied in practice to any lot because it is very difficult to ensure bulk representativeness of the true lot properties. What about the secondary sampling steps, i.e., all those necessary to produce the analytical sample from the bulk sample? Fortunately, the technical problems associated with all the secondary sampling steps are much easier to handle. Specifically, if grinding and mixing are properly carried out, the assumption of random distribution of GM material in the "populations" sampled at each secondary sampling step (bulk, laboratory, and test samples) is statistically acceptable.

Let us consider an example (after Lischer, 2001b) that illustrates why this is the case. For unprocessed soybeans and maize kernels, analytical samples fully representative of 2.5 kg bulk samples can be produced by grinding the whole sample to sufficiently fine powder, mixing thoroughly, and taking 1 g samples for analysis. However, if the opposite sequence was followed, that is if 1 g from 2.5 kg sample was weighed out and then ground, the resulting sample would hardly be representative. Why? Because a 2.5 kg soybean sample, contains about 10,000 individual beans. When ground to fine powder, each bean is reduced to many thousands of fragments. When the powder from all beans in the 2.5 kg sample is mixed thoroughly, each 1 g sample of that mixture contains approximately an equal number of fragments from every bean in the 2.5 kg sample. On the other hand, if 1g of the original sample was weighted out for the analytical sample, it would contain only four or five of the 10,000 beans present in the original sample.

In summary, it is clear that assuming random distribution of GM material within lots to produce bulk samples presents a serious risk, because it encourages solutions to sampling problems that overlook the issue of heterogeneity (Lischer, 2001a). Nevertheless, the same assumption is acceptable in all the following secondary sampling steps involved in routine lot sampling if mixing, grinding, and dividing procedures are properly applied.

SAMPLING FOR GMO DETECTION/QUANTIFICATION: RECOMMENDATIONS

Providing recommendations for sampling approaches suitable for the detection/quantification of GM materials in different market products is a challenge. On one hand, a high likelihood of nonrandom distribution of GM contaminations in most market products is expected. On the other, experimental data regarding the distribution of GM materials in different products are lacking. Yet we know from sampling theory that the distribution of a contaminant in a bulk mass greatly affects the effectiveness of sampling procedures.

A necessary first step to address this problem is to estimate the extent to which heterogeneity affects the accuracy of the detection and/ or quantification of low levels of GM materials in large quantities of bulk commodities. In other words, it is important to know what errors could be made if GMO distribution assumptions were violated. It is clear that an approach free of the constraint implicit in the assumption of random distribution is necessary. Nevertheless, defining such an approach is a challenging task. The number of possible distributions of GMOs in the case of lot heterogeneity is impossible to define because it depends on three factors, all equally difficult to measure for any given lot: the number of GM contamination sources, their degree of spatial aggregation, and the level of contamination due to each source. At present no model of known properties can define such distributions a priori, thus making the theoretical definition of optimal sampling techniques impossible.

In order to address this issue, Paoletti et al. (2003) developed a software program, KeSTE (Kernel Sampling Technique Evaluation), designed to evaluate the efficiency of different sampling plans as a function of specific combinations of lot properties. The program is based on a two-step procedure. First, lots with user-defined properties (e.g., size of the lot, percentage of GMOs, number of sources of GMOs, and level of spatial aggregation of GMOs) are created, so that several instances of a lot can be simulated differently because of the constraints imposed on the distribution of GM material. Second, lots are sampled according to one or several sampling schemes of interest to the user. Response surfaces are built to identify proper sampling techniques as a function of the defined lot characteristics. Clearly, this software is designed to provide statistically reliable estimates of the sampling error associated to the first sampling step, the creation of the bulk sample, under different lot heterogeneity scenarios. It does not provide information regarding the secondary sampling steps. Indeed, as explained previously, the production of the bulk sample is the sampling operation that affects the magnitude of the overall sampling error associated to any given sampling protocols because of the specific difficulties of ensuring its representativeness of the *true* lot properties.

A first application of this software allowed the efficacy of different sampling strategies on different kernel lot heterogeneity scenarios to be investigated (Paoletti et al., 2003). The results are particularly interesting. It appears that (1) current procedures for the procurement of bulk samples, as stipulated in international guidelines (Table 3.1), are sensitive to nonrandom distribution of GM materials; (2) in cases of modest levels of lot heterogeneity, bulk samples have a high probability of not correctly representing the lot. Without a doubt, these exploratory results issue a clear warning with respect to the unconditional acceptance of standardized sampling procedures in absence of the knowledge of GM material distribution in kernel lots.

Once more, the clear need for verifying the distribution of GM materials in real lots is underlined. KeSTE is the first tool developed to identify proper sampling strategies as a function of specific lot properties, as long as they are known or can be predicted with a certain degree of certainty. This seems trivial. Yet the application of sampling theory to GMO analysis is limited precisely because of the complete absence of experimental data indicating GMO distribution in bulk commodities. As discussed extensively in this chapter, the definition of suitable sampling protocols is impossible without at least some information on population properties.

CONCLUDING REMARKS

A number of factors must be taken into account when defining sampling protocols. Among these, the definition of a maximum acceptable sampling error is of utmost importance. In the case of GMO analysis, the degree of risk that both the consumer and the producer are prepared to accept, in terms of getting a wrong result, will contribute to the definition of the maximum acceptable sampling error. Once this is defined, the sampling protocol can be designed accordingly, so that the sampling survey's costs can be minimized without compromising result reliability beyond a certain level (i.e., the accepted risk).

Nevertheless, when sampling is executed to check for compliance with legislation requirements (i.e., regulatory sampling), it is of crucial importance to ensure a high degree of confidence that the survey is accurate and that the sampling error is as small as possible. Specifically, if a legal threshold limit is set for acceptance of the presence of GM material (as is currently the case within the European Union), all adopted sampling protocols must ensure that such a threshold is respected with a certain degree of confidence. Of course, the lower this limit is, the greater the demands will be upon the sampling plans.

As far as GMO surveys are concerned, a series of documented information must be taken into account during the definition of sampling protocols. Results from both theoretical and simulation researches indicate that heterogeneity is much more likely to occur than homogeneity with respect to GMO distribution in bulk commodities. Data from experimental studies indicate that lots of bulk materials can show extensive heterogeneity for traits other than GMOs. Together, these results pose a serious limit to the unconditional acceptance of the assumption of random distribution of GM materials and to the use of binomial distribution to estimate producer and consumer risks.

So, where do we go from here? If providing reliable sampling recommendations is a priority for the scientific community, it is necessary to invest in research projects designed to collect data on the real distribution of GM materials worldwide. This would allow a proper calibration of the statistical models used to estimate the degree of expected lot heterogeneity. Being in a position to make statistically solid predictions regarding possible heterogeneity scenarios is a prerequisite to the definition of suitable sampling plans.

Meanwhile, some precautions should be taken. Because raw materials often come from different suppliers, and given that industrial activities are structured in space and time, we can expect a portion of the original chronological order to be present in the spatial structure of any lot. Under this assumption, a systematic sampling approach is recommended over a random sampling approach. Systematic sampling should take place when lots are being unloaded, i.e., when the option of continuous sampling of the entire consignment exists. This is preferable to the situation of large batches in silos or trucks, where it is difficult to access remote parts even when employing specialized equipment such as sampling probes.

As far as the number of increments used to produce the bulk sample is concerned, it is difficult to make clear recommendations because the number of increments required to minimize the sampling error, according to some predefined expectation, will depend entirely upon the heterogeneity of the lot under investigation. The complete lack of data on the expected distribution of real lots makes it impossible to establish objective *criteria* to address this problem. Nonetheless, when defining the number of increments to be sampled, the results from the simulation study (Paoletti et al., 2003), indicating that even modest levels of heterogeneity are going to compromise sampling reliability when 30 to 50 increments are used to produce the bulk sample, should be taken into account.

Following the secondary sampling steps necessary to produce final samples of suitable working size from large bulk samples will always be necessary. Nevertheless, assuming random distribution to estimate the errors associated to each of these secondary sampling steps is not going to be a problem, as long as handling of the material, in terms of both grinding and mixing, is properly carried out. Sample size reduction should be attempted only when all of the sampled material is reduced to the same particle size: the smaller and the more uniform material is after milling, the more successful mixing will be to ensure homogeneity in the population and, ultimately, to minimize the sampling error.

Unfortunately, correct sampling is often completely uncorrelated with sampling costs: a protocol structured according to the recommendations listed earlier will have a high cost in terms of both time and financial resources necessary to complete the sampling operation. Nevertheless, excuses to perform incorrect sampling cannot be justified by time and money limitations. A correct sampling is always accurate. If not, no reason exists to carry out any sampling at all.

REFERENCES

Belton, P. (2001). Chance, risk, uncertainty, and food. *Food Sci. and Technol.* 12:32-35.

Bowes, D.M. (1997) The new green revolution: Transgenic crops—Panacea or panic: A consumer overview of some recent literature. In *Commercialisation of Transgenic Crops: Risk, Benefit, and Trade Considerations* (pp. 363-373), G.D. McLean, P.M. Waterhouse, G. Evans, and M.J. Gibbs, eds., Australian Government Publishing Service, Canberra.

Cochran, G.W. (1977). *Sampling Techniques*, Third Edition. J. Wiley and Sons Inc., New York.

Colwell, R., Norse, E.A., Pimentel, D., Sharples, F.E. and Simberoff, D. (1985). Genetic engineering in agriculture. *Science* 229:111-112.

Comité Européen de Normalisation (CEN) (2001). CEN/TC 275/WG 11 N 111. Detection of genetically modified organisms and derived products—sampling, draft, January.

Copeland, L.O., Liu, H., and Schabenberger, O. (1999). Statistical aspects of seed testing. *Seed Technol. Newslet.* 73(1):20.

Coster, R. (1993). Seed lot size limitations as reflected in heterogeneity testing. *Seed Sci. and Technol.* 21:513-520.

Dale, P.J. (1994). The impact of hybrids between genetically modified crop plants and their related species: General considerations. *Molec. Ecol.* 3:31-36.

Dale, P.J., Clarke, B., and Fontes, E.M.G. (2002). Potential for the environmental impact of transgenic crops. *Nat. Biotechnol.* 20:567-574.

Darmancy, H. (1994). The impact of hybrids between genetically modified crop plants and their related species: Introgression and weediness. *Molec. Ecol.* 3:37-40.

Darmancy, H., Lefol, E., and Fleury, A. (1998). Spontaneous hybridizations between oilseed rape and wild radish. *Molec. Ecol.* 7:1467-1473.

Ellstrand, N.C. (1993). How ya gonna keep transgenes down on the farm? *The Amicus Journal* 15:31.

Ellstrand, N.C. and Elam, D.R. (1993). Population genetic consequences of small population size: Implication for plant conservation. *Ann. Rev. Ecol. Syst.* 24: 217-242.

Ellstrand, N.C. and Hoffman, C.A. (1990). Hybridization as an avenue of escape for engineered genes. *Bioscience* 40:438-442.

Ellstrand, N.C., Prentice, H.C., and Hancock, J.F. (1999). Gene flow and introgression from domesticated plants into their wild relatives. *Ann. Rev. Ecol. Syst.* 30:539-563.

European Commission (EC) (1998). Commission Directive 98/53/EC of 16 July 1998 laying down the sampling methods and the methods of analysis for the official control of the levels for certain contaminants in foodstuffs. <http://europa. eu.int/eurolex/en/lif/dat/1998/en398L0053.html>.

Food and Agriculture Organization/World Health Organization (FAO/WHO) (1991). *Strategies for Assessing the Safety of Foods Produced by Biotechnology.* Joint FAO/WHO Consultation, Genéve, Swizerland.

Food and Agriculture Organization/World Health Organization (FAO/WHO) (2001). *Safety Aspects of Genetically Modified Foods of Plan Origin.* Report of a joint FAO/WHO Expert Consultation, Switzerland, June 2000.

Gilbert, J. (1999). Sampling of raw materials and processed foods for the presence of GMOs. *Food Control* 10:363-365.

Hübner, P., Waiblinger, H.-U., Pietsch, K., and Brodmann, P. (2001). Validation of PCR methods for the quantification of genetically modified plants in food. *J. AOAC Int.* 84:1855-1864.

International Organization for Standardization (ISO) (1990). *Oilseeds—Sampling.* ISO-542. ISO, Genéve, Swizerland.

International Organization for Standardization (ISO) (1995). *Sampling Procedures for Inspection By Attributes. Part 0: Introduction to the ISO-2859 Attributed Sampling System.* ISO-2859-0. ISO, Genéve, Swizerland.

International Organization for Standardization (ISO) (1999). *Cereals, Pulses and Milled Products—Sampling of Static Batches.* ISO-13690. ISO, Genéve, Swizerland.

International Seed Testing Association (ISTA) (1986). *Handbook on Seed Sampling.* ISTA, Zurich, Swizerland.

International Seed Testing Association (ISTA) (2003). *International Rules for Seed Testing.* ISTA, Zurich, Swizerland.

Jorgensen, J. and Kristensen, K. (1990). Heterogeneity of grass seed lots. *Seed Sci. Technol.* 18:515-523.

Kay, S. (2002). Comparison of sampling approaches for grain lots. *European Commission Report,* Code EUR20134EN, Joint Research Centre Publication Office.

Kay, S. and Paoletti, C. (2002). Sampling strategies for GMO detection and/or quantification. *European Commission Report,* Code EUR20239EN, Joint Research Centre Publication Office.

Krawczyk, T. (1999). Resolving regulatory rifts: European and US GMO systems generate controversy. *Inform* 10(8):756-767.

Kruse, M. (1997). The effects of seed sampling intensity on the representativeness of the submitted sample as depending on heterogeneity of the seed lot. *Agrobiol. Res.* 50:128-145.

Kruse, M. and Steiner, A.M. (1995). *Variation Between Seed Lots As an Estimate for the Risk of Heterogeneity with Increasing ISTA Maximum Lot Size* (Abstract 21). Presented at ISTA Twenty-Fourth Congress Copenhagen, Seed Symposium.

Kuiper, A.H., Kleter, A.G., Noteborn, P.J.M.H., and Kok, J.E. (2001). Assessment of the food safety issues related to genetically modified foods. *Plant J.* 27 (6):503-528.

Lischer, P. (2001a). Sampling procedures to determine the proportion of genetically modified organisms in raw materials. Part I: Correct sampling, good sampling practice. *Mitt. Lebensm. Hyg.* 92: 290-304.

Lischer, P. (2001b). Sampling procedures to determine the proportion of genetically modified organisms in raw materials. Part II: Sampling from batches of grain. *Mitt. Lebensm. Hyg.* 92:305-311.

Paoletti, C., Donatelli, M., Kay, S., Van den Eede, G. (2003). Simulating kernel lot sampling: The effect of heterogeneity on the detection of GMO contaminations. *Seed Sci. Technol.* 31(3):629-638.

Philipon, P. (1999). Aliments transgéniques: Le doute s'installe. *Biofutur* 192:16-20.

Remund, K., Dixon, D., Wright, D., and Holden, L. (2001). Statistical considerations in seed purity testing for transgenic traits. *Seed Sci. Res.* 11:101-120.

Simmonds, N.W. (1993). Introgression and incorporation: Strategies for the use of crop genetic resources. *Biol. Rev.* 68:539-562.

Slatkin, M. (1985). Gene flow in natural populations. *Ann. Rev. Ecol. Syst.* 16:393-430.

Slatkin, M. (1987). Gene flow and the geographic structure of natural populations. *Science* 236:787-792.

Snow, A.A. (2002). Transgenic crops—Why gene flow matters. *Nat. Biotechnol.* 20:542-541.

Tattersfield, J.G. (1977). Further estimates of heterogeneity in seed lots. Seed Science and Technology, 5:443-450.

U.S. Department of Agriculture, Grain Inspection, Packers and Stockyards Administration (USDA GIPSA) (1995). *Grain Inspection Handbook*, Book 1. USDA GIPSA, Washington DC. <http://www.usda.gov/gipsa/reference-library/handbooks/grain-insp/grbook1/bk1.pdf>.

Van Raamsdonk, L.W.D. (1995). The effect of domestication on plant evolution. *Acta Botanica Neerl*, 44(4):421-438.

World Health Organization and the Food and Agriculture Organization of the United Nations (WHO-FAO) (2001). Proposed draft general guidelines on sampling (CX/MAS 01/3). Codex Alimentarius Commission, Codex Committee on Methods of Analysis and Sampling.

Chapter 4

Reference Materials and Standards

Stefanie Trapmann
Philippe Corbisier
Heinz Schimmel

INTRODUCTION

Certified reference materials (CRMs) are essential tools in quality assurance of analytical measurements. CRMs help to improve precision and trueness of analytical measurements and ensure the comparability of achieved results. CRMs are designed mostly for the development and validation of new methods and the verification of correct application of standardized methods and trueness of results, but can also be developed for calibration purposes. CRMs are produced, certified, and used in accordance with relevant International Organization for Standardization (ISO) and Community Bureau of Reference (BCR) guidelines (ISO, 2000; BCR, 1994, 1997). In order to prevent variation and drifting of the measuring system, homogeneity and stability of a CRM are of utmost importance. Precise knowledge about the used base material and the influence of the different production steps on the analyte is needed to reproduce CRM batches.

Compared to reference materials for well-defined molecules of low molecular mass, the complexity of DNA and protein molecules requires the consideration of additional parameters, such as structural data and degradation rate. Essentially, reference materials for macromolecules such as DNA and proteins have to behave similar to routine samples to ensure full applicability and optimal impact with regard to standardization of measurement procedures applied in the field. As in clinical chemistry, these phenomena can be summarized under the term *commutability* (Schimmel et al., 2002).

In general, CRMs are designed for specific intended use, and possible fields of application are laid down in the certificate. The fields of use are subject to previous investigations to ensure that the CRM is suitable for a given purpose. Various materials can be used as CRMs for the detection of genetically modified organisms (GMOs), matrix-based materials produced (e.g., from seeds), or pure DNA and protein standards. Pure DNA standards can either consist of genomic DNA (gDNA) extracted, for example, from plant leaves, or of plasmidic DNA (pDNA) in which a short sequence of a few hundred base pairs (bp) has been cloned, and which has been amplified in a suitable bacterial vector. Pure proteins can be extracted from ground seed or produced using recombinant technology.

Depending on their use, for identification and/or quantitation of GMOs, different requirements have to be met by the CRMs. The various types of GMO CRMs have specific advantages and disadvantages. In general, the measurement of the GMO content and the related production of GMO CRMs for verification and calibration purposes can be considered as a special case because the weight percentage of GMOs present in a sample cannot be measured directly. The first GMO CRMs made available by the Joint Research Centre (JRC) of the European Commission were powdery, matrix-based CRMs produced from GMO and non-GMO kernels. They have been used for DNA- and protein-based methods, determining either the transgenic DNA sequence or the expressed transgenic protein typical for a GMO. Due to their uniqueness, these CRMs have been used not only for the control of GMO detection methods but also for the calibration of such measurements.

MATRIX-BASED GMO CRMs

Since the late 1990s, three generations of GMO CRMs for maize and soybeans have been produced, and each generation reflects major improvement of the production procedure. Up to now, different weight percentages in the range of 0 to 5 percent GMO for four GM events authorized in Europe have been produced and certified: Bt (*Bacillus thuringiensis*)-11 and Bt-176 maize, Roundup Ready soybean, and Monsanto (MON) 810 maize (Table 4.1) (IRMM, 2002). The production of powdery GMO CRMs includes several production steps, the major ones being (Figure 4.1) as follows:

- Characterization of the kernels
- Grinding of the kernels
- Mixing of different weight portions
- Bottling and labeling
- Control of the final product and certification

Characterization of the used base materials is the first production step, and it is of high importance especially in the field of GMO CRMs.

TABLE 4.1. Produced and certified matrix-based GMO CRMs

GMO CRM[a] (production year)	GM event	Raw materials used (GMO/non-GMO material)[c]	Production technique
1st generation			
IRMM-410 (1997/1998)	Roundup Ready soy	Homozygous Roundup Ready soy/biological framed grains	Single-step grinding,wet-mixing (Pauwels et al., 1999a,b)
IRMM-411 (1998/1999)	Bt-176 maize	Heterozygous Furio CB-XRCM/Furio F0491 grains	
IRMM-412 (1999)	Bt-11 maize	Heterozygous Bt-176/Furio F0491 grains	
2nd generation			
IRMM-410R (1999/2000)	Roundup Ready soy	Homozygous High Cycle Brand 9988.21.RR soy/T 77 seeds	Single-step grinding, wet-mixing with cooling (Trapmann, Le Guern, et al., 2000)
3rd generation			
IRMM-413 (2000/2001)	MON 810 maize	Heterozygous cultivar DK 513/near isogenic DK 512 non-trade-variety seeds	Two-step grinding with cryo-grinding, dry-mixing (Trapmann, Le Guern, et al., 2001)
IRMM-410S (2001)	Roundup Ready soy	Homozygous RR line AG5602 RR/AG line A1900 seeds	Single-step grinding, dry-mixing, partial second-step grinding (Trapmann, Catalani, et al., 2002)
IRMM-412R[b] (2002)	Bt-11 maize	Heterozygous NX3707 Bt-11/near isogenic Pelican seeds	Two-step grinding, dry-mixing
IRMM-411R[b] (2002)	Bt-176 maize	Heterozygous Garona Bt-176/near isogenic Bahia seeds	

[a] Replacement batches indicated with a capital R or S
[b] No distribution agreement
[c] Hetero- and homozygocity in respect of the GM event

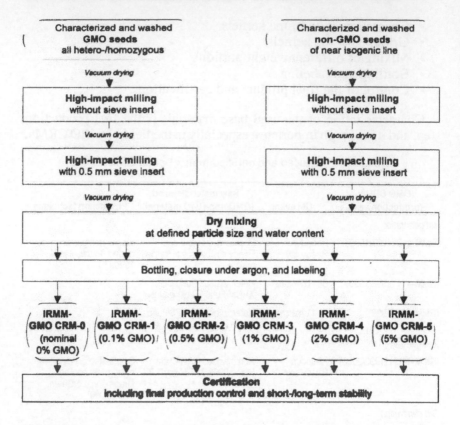

FIGURE 4.1. Production steps for powdery, third-generation maize matrix-based GMO CRMs

GMO and non-GMO base materials of seed quality, representing only one variety, permits knowledge about its genetic composition. The base materials are analyzed for their purity after cleaning the kernels with demineralized water in order to minimize carryover contamination. Furthermore, the portion of homozygous and heterozygous kernels in the GMO base materials has to be determined in order to allow the later use of the CRM for quantitative analysis. Factors such as the zygosity of a genome influence the results of the indirect GMO quantitation with DNA-based detection methods. For those reasons, a near isogenic parental line is preferably chosen as a non-GMO base material (Van den Eede et al., 2002).

During production of the third generation of matrix-based GMO CRMs, it proved difficult to obtain non-GMO raw materials that were completely free of GM contamination. The contamination found in the non-GMO raw material used for the production of Institute for Reference Materials and Measurements (IRMM)-410S corresponds to a less than 0.02 percent genetic modification (Table 4.2). IRMM-410S-0 was therefore certified nominally as 0 percent, containing less than 0.03 percent GMO. A similar observation was made in the case of MON 810, which was certified as containing less than 0.02 percent GMO.

In addition, both base materials have to be analyzed for their total DNA content to allow the later use of these CRMs for DNA-based analysis. Because determination methods for the total DNA content focus on different parts of the DNA different methods are applied preferably at the same time. After extraction with perchloric acid, spectrophotometric methods measure 2-deoxyribose. After digestion with formic acid, liquid chromatography measures thymine, and after digestion with NaOH, fluorometric methods measure nucleic acids (Table 4.3) (Trapmann, Schimmel, et al., 2002). Special attention should be paid to the extraction step. The extraction efficiency of methods applied has to be the same for the GMO material as for the non-GMO material in order to obtain correct quantitation results.

TABLE 4.2. Certification of nominal 0 percent IRMM-410S (Roundup Ready soybean)

Number of analysis (n)	EPSPS event-specific rt-PCR[a]	NOS specific rt-PCR	CaMV 35S specific rt-PCR[a]
(n)	GM detected (%)	GM detected (%)	GM detected (%)
3	0.011 ± 0.009	0.004 ± 0.004	0.006 ± 0.002
3	0.008 ± 0.007	0.006 ± 0.003	0.008 ± 0.007
3	0.030 ± 0.014	0.013 ± 0.009	0.005 ± 0.002
3	0.017 ± 0.010	0.009 ± 0.003	0.004 ± 0.002
3	0.006 ± 0.004	0.005 ± 0.003	0.005 ± 0.003
Average	**0.015 ± 0.012**	**0.007 ± 0.005**	**0.006 ± 0.003**
LOD (3.3 σ/S)	0.02%	0.01%	Low specificity

Source: Trapmann, Corbisier, et al., 2002.
[a]according to Bundesamt für Gesundheit

TABLE 4.3. Determination of the total DNA content in GMO and non-GMO raw materials

Analyte (detection and extraction method)	2-deoxyribose (spectrophotometric after extraction with perchloric acid)	Thymine (HPLC after digestion with formic acid)	Nucleic acids (fluorometric after digestion with NaOH or magnetic bead extraction)
Bt-176 maize (IRMM-411)	1.0 ± 0.1	0.9 ± 0.1	1.2 ± 0.2[a]
RR soy (IRMM-410/410R)	1.0 ± 0.1	1.0 ± 0.1	1.0 ± 0.3[a]
MON 810 maize (IRMM-413)	1.0 ± 0.1	1.0 ± 0.2	1.4 ± 0.3[b]

A crucial point is the calculation of the optimal particle size for the production of a homogenous GMO CRM with an acceptable certainty. Larger particles lead to an increased contribution to the uncertainty of a CRM because the different tissue types of the kernels contain not only different amounts of DNA (due to the cell size), but also different sets of genomes (due to embryogenesis) (Prokisch, Zeleny, et al., 2001). Too small particle size often leads to a sticky powder unsuitable for a dry-mixing procedure (Trapmann, Schimmel, et al., 2001). The optimal average particle size for different matrixes can differ enormously and has to be studied prior to a production with the help of homogeneity studies (Trapmann, Le Guern, et al., 2001; Trapmann, Catalani, et al., 2002).

In addition to the optimal particle size distribution for the production of GMO CRMs, one has to consider that particle size has a direct and indirect impact on the analytical results. For example, it has been observed that particle size can influence directly the result of protein-based detection methods. Fourteen to 24 percent more MON 810 were detected by laboratories participating in an international collaborative trial when using the finer ground CRM IRMM-413 (average particle size 35 μm) in comparison to a material produced from the same raw material with an average particle size of approximately 150 μm. It was concluded that the results of the differently ground raw materials led to a different protein extractability for the applied method

(Stave et al., 2000). The indirect impact is linked to the number of particles in a sample. In order to have 100 GM particles of an estimated density of 0.8 in a 1 percent GMO sample, 34 g of a laboratory sample (with a typical average particle size of 2,000 μm) would be needed, whereas only 8 mg (average particle size 125 μm) would be needed from a typical GMO CRM.

Matrix-based GMO CRMs are produced by mixing GMO and non-GMO powders on a weight/weight basis. Both powders are therefore corrected for their water content prior to the mixing procedure. A dry-mixing technique has been developed ensuring homogeneous mixing of the GMO powders with the non-GMO powders, and at the same time avoiding degradation of the DNA analyte (Trapmann, Schimmel, Le Guern, et al., 2000). During dry-mixing the powders are diluted stepwise.

The homogeneity of matrix-based GMO CRMs mixed by a dry-mixing technique was controlled in a study carried out with Au-spiked powders. Different concentration levels were prepared by stepwise dilution of the 100 percent Au-spiked powder. The Au content was then determined with the help of neutron activation analysis (NAA). NAA was performed at a sample intake level of 50 mg, which is common for PCR measurements (Table 4.2). Homogeneity can be ensured only if the powders mixed have similar average and maximum particle size and similar water content as those used in a successful homogeneity study. As pointed out previously, the results cannot be transferred from one matrix type to the other. The powders are vacuum dried at temperatures below 30°C in order to avoid affecting the protein analyte.

After bottling, closure of the bottles under argon and labeling the control of the production and control of the final product is assured with the help of particle-size distribution analysis, water content determination, and water activity determination. The status of the analyte during production and in the final product is generally controlled by gel electrophoresis and DNA- and protein-based detection methods.

A complete uncertainty budget calculation is made during the certification. Contributing uncertainties are the uncertainties due to the weighing procedure, the water content determination, the inhomogeneity at the recommended sample intake level (for the GMO CRMs, commonly 100 mg), and the purity of the GMO and non-GMO base materials (Ellison et al., 2000).

Thorough control of homogeneity and stability are essential for the certification of reference materials. Control ensures the validity of the certificate for each bottle of a batch throughout a defined shelf life (Pauwels et al., 1998; Lamberty et al., 1998). Inhomogeneity or instability will lead to the failure of method validations and may lead to inaccurate conclusions.

After successful certification, the stability of matrix GMO CRMs is controlled in six-month intervals using gel electrophoresis, methods for the determination of the total DNA content, and DNA- and protein-based detection methods. For the first generation GMO CRMs, a shelf life of 42 months at 4°C has been determined. After that time, DNA fragmentation did not allow the application of DNA-based methods targeting amplicons of 195 bp anymore. Unfortunately, DNA-based detection methods do not give information about the complete gDNA of a CRM, but only about the sequence targeted. For that reason, total DNA measurements and gel electrophoresis are carried out during stability monitoring. However, third-generation GMO CRMs might show a longer shelf life due to the different production techniques used.

The user of a matrix-based GMO CRM has to keep in mind that DNA and protein degradation is likely to occur during food production especially in the case of homogenous CRMs. Due to earlier indications (Trapmann, Schimmel, Brodmann, et al., 2000), two studies have been carried out to determine whether degradation of soybean and maize flour can influence quantitation with DNA- and protein-based detection methods. Roundup Ready soybean and MON 810 maize powder CRMs with a GMO concentration of 1 percent were subjected to various degrees of degradation. The CRM materials were exposed to water at different temperature levels and to different shear forces. The fragmentation degree was then characterized by gel and capillary electrophoresis and the percentage of genetically modified soybean was determined by double quantitative real-time polymerase chain reaction (RT-PCR) and ELISA (enzyme-linked immunosorbent assay). The GMO content in the Roundup Ready soybean sample exposed to higher temperatures (40-45°C) and high shear forces could not be measured correctly by RT-PCR. Four times higher concentrations of GMO were measured, whereas the application of an ELISA Bt soy test kit led to the correct quantitation (Table 4.4). The DNA amounts per PCR reaction were kept on the same level. No such effects have been found on the MON 810 samples (Corbisier et al., 2002).

TABLE 4.4. Soybean degradation study (RT-PCR was performed on 50 ng DNA extracted from 20 mg powder)

Treatment IRMM-410S-3 (1% RR soy)	Number of extractions (n)	Lectin RT-PCR[a] (%)	Epsps RT-PCR[a] (%)	Normalized GM concentration (%)
Control (without water exposure)	5	94.62 ± 9.42	0.67 ± 0.07	1.00 ± 0.11
Gently stirred, 4°C	5	102.26 ± 7.81	0.98 ± 0.18	1.13 ± 0.12
Propeller mixed, 4°C	5	66.48 ± 3.76	0.59 ± 0.11	1.25 ± 0.20
Turrax mixed, 4°C	5	64.00 ± 5.46	0.53 ± 0.06	1.17 ± 0.05
Propeller mixed, 40-45°C	5	78.88 ± 15.72	0.54 ± 0.09	0.97 ± 0.11
Turrax mixed, 40-45°C	10	94.41 ± 4.42	2.83 ± 0.36	4.20 ± 0.45
Baked, 250°C	5	74.26 ± 18.80	0.72 ± 0.17	1.38 ± 0.09

[a]according to Bundesamt für Gesundheit

PURE GENOMIC DNA GMO CRMs

Stability and applicability of gDNA CRMs for PCR methods has been shown successfully in the frame of comparable projects. For the detection of foodborne pathogens, dried gDNA standards have been produced, each vial containing 1-2 µg DNA, which could be recovered easily by addition of water. No differences were observed in the amplification between the dried standards and the raw material (Prokisch, Trapmann, et al., 2001). One has to be aware that GM patent holders have to give approval for the production and use of gDNA.

PURE PLASMID DNA GMO CRMs

DNA fragments containing a GMO-specific region or a single-copy endogenous gene can be cloned in a high copy number plasmid vector. This plasmid is transformed into competent bacterial strains and extracted and purified from a bacterial culture. The plasmid DNA concentration is then quantified. Since the precise size of the vector is known, the copy numbers can be determined and diluted to obtain, e.g., 10^8 to 10^0 copies of each fragment. Calibration curves generated with those plasmids can be used in RT-PCR experiments by plotting the threshold cycle (C_T) versus the starting quantity of plasmid. This approach was demonstrated by Taverniers et al. (2001) for the detection of the Roundup Ready soybean. Two plasmids, pAS104 and pAS106, containing a 118 bp fragment of the lectin *Le1* gene and a 359 bp fragment of the MaV 35S promoter/plant junction were cloned and are available at the BCCMTM/LMBP collection (under accession numbers LMBP 4356 and LMBP 4357 respectively). Kuribara et al. (2002) followed the same strategy and have cloned DNA fragments of five lines of genetically modified maize, including MON 810, Bt-11, Bt-176, T25, and GA21 in a single plasmid (pMuI5), as well as a Roundup Ready soybean (pMiISL2). Those two plasmids were successfully used as calibrants for the qualitative and quantitative PCR techniques (Kuribara et al., 2002). Kuribara et al. (2002) experimentally determined a coefficient value (C_V), which represents the ratio between recombinant DNA and the endogenous sequence in seeds. The theoretical coefficient value of a single copy of transgene per genome is expected to be 1 for homozygous material and less than 1 for heterozygous materials, depending on the ploidy level. Kuribara et al. (2002) observed differences between the experimental and theoretical values that could be attributed to differences of PCR efficiencies that resulted from the presence of nontargeted background DNA in the case of gDNA. Roche Diagnostics is also commercializing two detection kits for the quantitative detection of Roundup Ready soybean and Bt-11 using pDNA as a calibrator. The internal DNA calibrator is provided with a correction factor to compensate batch differences in the production of the plasmids.

The commutability of the pDNA CRMs has to be investigated if the materials are going to be used for quantitation of genomic DNA. For example, pDNA may behave different from gDNA because it is normally significantly shorter, and relatively small differences of the amplification efficiency during the PCR reaction have a pronounced effect on the quantitative result because of the exponential nature of the reaction (Schimmel et al., 2002). In addition, one has to be aware that the patents of some GM events cover DNA fragments as well.

PURE PROTEIN GMO CRMs

Reference materials are also very important for protein detection and quantitation. In that regard, not only are the properties of the measure relevant, but also those of the matrix. In most cases, proteins have to be characterized in a multiparametric approach to ensure fitness for purpose of the reference materials produced. The relevant parameters have to be defined case by case because of the large biological variability of proteins and differences in measurement procedures applied in the field (Schimmel et al., 2002).

The protein expression level differs in the different parts of a plant and depends largely on the growing conditions. The transgenic Cry1Ab protein can, for instance, be found only in negligible concentrations in Bt-176 maize seeds, whereas it is expressed in the seeds of Bt-11 maize, although the concentrations are even higher in leaf material (Stave, 2002). Consequently, it is difficult to translate quantitative results obtained with the help of pure protein GMO CRMs into a GMO concentration unless working on kernels and using sampling theories (USDA, 2002). Some events, such as GA21 maize, cannot be specifically detected without the help of protein-based detection methods.

DISCUSSION

Table 4.5 summarizes the specific advantages and disadvantages of the various types of GMO CRMs. Matrix-based GMO CRMs allow the analytical measurement procedure to be controlled, and help to determine extraction efficiency, matrix effects, and degradation of the analyte. Furthermore, matrix-based GMO CRMs can be used for DNA-based and protein-based determination methods, and help to ensure the comparability of both. On the other hand, DNA and protein content have the potential to vary naturally in concentration in different batches of raw materials, and their production is very time-consuming. However, the availability of raw material for the production of matrix-based GMO CRMs is restricted. The use of seeds for any other purpose than planting of crops is prohibited. Therefore, agreement contracts have to be made prior to the production of a matrix-based GMO CRM. The variation of the genetic background is a further disadvantage of matrix DNA and gDNA CRMs, since DNA-

TABLE 4.5. Advantages and disadvantages of various types of GMO CRMs

Type of CRM	Advantages	Disadvantages
Matrix-based GMO CRMs	Suitable for protein- and DNA-based methods	Different extractability (?)
		Large production needed
	Extraction covered	Low availability of raw material due to restricted use of seeds
	Commutability	
		Degradation
		Variation of the genetic background
Genomic DNA CRMs	Good calibrant	Large production needed
	Fewer seeds needed	Low availability of raw material due to restricted use of seeds
	Commutability	
		Variation of the genetic background
		Long-term stability (?)
Pure protein CRMs	Fewer seeds needed	Commutability (?)
Plasmidic DNA CRMs	Easy to produce in large quantities	Plasmid topology
		Discrepancies
	Broad dynamic range	Commutability (?)

based GMO quantitation may depend on varieties. Therefore one has to be careful to draw quantitative conclusions from measurements of unknown samples containing other varieties. Matrix-based GMO CRMs proved to be valuable tools for the validation of qualitative PCR in screening for 35S promoters and nopaline synthase (NOS) terminators (Lipp et al., 1999; Lipp et al., 2000). In the meantime, GMO CRMs have been used successfully in various quantitative validation studies, such as the CP4 EPSPS protein ELISA (measuring the 5-enolpyruvylshikimate-3-phosphate synthase protein derived from *Agrobacterium* sp. strain CP4) (Stave et al., 2000). A validation study for an event-specific RT-PCR method on MON 810 is currently being organized by the Federal Institute for Risk Assessment (Berlin), Joint Research Center of the European Commission, American Association of Cereal Chemists, and GeneScan. It is common for the mentioned collaborative studies that the same powdery reference materials have been used for calibration and as blind samples. Consequently, the results show fairly good standard deviations. Furthermore, one must be aware that this approach risks overlooking certain critical aspects of GM quantitation.

As it is unlikely that in the near future non-GMO seed materials of a near isogenic parental line will be found containing no traces of GM events, it will be impossible to produce an actual 0 percent matrix-based GMO CRM. This problem could be overcome easily with pDNA, which is easier to produce on a large scale, less time-consuming, and less costly than gDNA extractions, since femtograms of pDNA would be needed compared to the milligrams of gDNA required for a PCR reaction. However, some fundamental aspects are not yet understood, such as the topology of the reference plasmids (linearized vs circular/ supercoiled), the stability and precision of extreme dilutions (femtograms in microliters), the absence of PCR inhibitors in plasmids compared to extracted gDNA, the putative differential PCR amplification efficiencies between pDNA and gDNA, or the variable ratio between transgenes and endogenes in GM seeds.

REFERENCES

Bundesamt für Gesundheit (2003). *Swiss Food Manual (Schweizerisches Lebensmittelbuch)*. Eidgenössische Drucksachen- und Materialzentrale, Bern, Switzerland.

Community Bureau of Reference (BCR) (1994). *Guideline for the Production and Certification of BCR Reference Materials*, Document BCR/48/93. European Commission, DG Science, Research and Development, Brussels, Belgium.

Community Bureau of Reference (BCR) (1997). *Guidelines for the Production and Certification of BCR Reference Materials*, Document BCR/01/97—Part A. European Commission, DG Science, Research and Development, Brussels, Belgium.

Corbisier, P., Trapmann, S., Gancberg, D., Van Iwaarden, P., Hannes, L., Catalani, P., Le Guern, L., and Schimmel, H. (2002). Effect of DNA fragmentation on the quantitation of GMO. *Eur. J. Biochem.* 269(Suppl 1):40.

Ellison, S.L.R., Rosslein, M., and Williams, A. (2000). *EURACHEM/CITAC Guide: Quantifying Uncertainty in Analytical Measurement*. EURACHEM, London, United Kingdom.

Institute for Reference Materials and Measurements (IRMM) (2002). *Reference Materials*. <http://www.irmm.jrc.be>.

International Organization for Standardization (ISO) (2000). *ISO Guide 34*. ISO, Geneva, Switzerland.

Kuribara, H., Shindo, Y., Matsuoka, T., Takubo, K., Futo, S., Aoki, N., Hirao, T., Akiyama, H., Goda, Y., Toyoda, M., and Hino, A. (2002). Novel reference molecules for quantitation of genetically modified maize and soybean. *J. AOAC Int.* 85(5):1077-1089.

Lamberty, A., Schimmel, H., and Pauwels, J. (1998). The study of the stability of reference materials by isochronous measurements. *Fresenius J. Anal. Chem.* 360:359-361.

Lipp, M., Anklam, E., Brodmann, P., Pietsch, K., and Pauwels, J. (1999). Results of an interlaboratory assessment of a screening method of genetically modified organisms in soy beans and maize. *Food Control* 10:379-383.

Lipp, M., Anklam, E., and Stave, J.W. (2000). Validation of an immunoassay for the detection and quantification of Roundup-Ready soybeans in food and food fractions by the use of reference materials. *J. AOAC Int.* 83:919-927.

Pauwels, J., Kramer, G.N., Schimmel, H., Anklam, E., Lipp, M., and Brodmann, P. (1999a). *The Certification of Reference Materials of Maize Powder with Different Mass Fractions of Bt-176 Maize*, EC certification report EUR 18684 EN. EurOP, Luxembourg, Belgium.

Pauwels, J., Kramer, G.N., Schimmel, H., Anklam, E., Lipp, M., and Brodmann, P. (1999b). *The Certification of Reference Materials of Soya Powder with Different Mass Fractions of Roundup Ready Soya*, EC certification report EUR 18683 EN. EurOP, Luxembourg, Belgium.

Pauwels, J., Lamberty, A., and Schimmel, H. (1998). Quantification of the expected shelf-life of certified reference materials. *Fresenius J. Anal. Chem.* 361:395-399.

Prokisch, J., Trapmann, S., Catalani, P., Zeleny, R., Schimmel, H., and Pauwels, J. (2001). Production of genomic DNA reference materials for the detection of pathogens in food, poster at 115th AOAC international annual meeting and Exposition, September 9-13, Kansas City, MO.

Prokisch, J., Zeleny, R., Trapmann, S., Le Guern, L., Schimmel, H., Kramer, G.N., and Pauwels, J. (2001). Estimation of the minimum uncertainty of DNA concentration in a genetically modified maize sample candidate certified reference material. *Fresenius J. Anal. Chem.* 370:935-939.

Schimmel, H., Corbisier, P., Klein, C., Philipp, W., and Trapmann, S. (2002). New reference materials for DNA and protein detection, lecture at 12th Euroanalysis, September 8-13, Dortmund, Denmark.

Stave, J.W. (2002). Protein immunoassay methods for detection of biotech crops: Application, limitations, and practical considerations. *J. AOAC Int.* 85(3):780-786.

Stave, J.W., Magin, K., Schimmel, H., Lawruk, T.S., Wehling, P., and Bridges, A. (2000). AACC collaborative study of a protein method for the detection of genetically modified corn. *Cereal Food World* 45:497-501.

Tavernier, I., Windels, P., Van Bockstaele, E., and De Loose, M. (2001). Use of cloned DNA fragments for event-specific quantification of genetically modified organisms in pure and mixed food products. *Eur. Food Res. Technol.* 213:417-424.

Trapmann, S., Catalani, P., Conneely, P., Corbisier, P., Gancberg, D., Hannes, E., Le Guern, L., Kramer, G.N., Prokisch, J., Robouch, P., Schimmel, H., Zeleny, R., Pauwels, J., Van den Eede, G., Weighardt, F., Mazzara, M., and Anklam, E. (2002). *The Certification of Reference Materials of Dry-Mixed Soya Powder with Different Mass Fractions of Roundup Ready Soya*, EC certification report EUR 20273 EN. EurOP, Luxembourg, Belgium.

Trapmann, S., Corbisier, P., Gancberg, D., Robouch, P., and Schimmel, H. (2002). Certification of "nominal 0 percent" GMO CRMs, poster at 116th AOAC International Annual Meeting and Exposition, September 22-26, Los Angeles, CA, USA.

Trapmann, S., Le Guern, L., Kramer, G.N., Schimmel, H., Pauwels, J., Anklam E., Van den Eede, and G., Brodmann, P. (2000). *The Certification of a New Set of Reference Materials of Soya Powder with Different Mass Fractions of Roundup Ready Soya*, EC certification report EUR 19573 EN. EurOP, Luxembourg, Belgium.

Trapmann, S., Le Guern, L., Prokisch, J., Robouch, P., Kramer, G.N., Schimmel, H., Pauwels, J., Querci, M., Van den Eede, G., and Anklam, E. (2001). *The Certification of Reference Materials of Dry Mixed Maize Powder with Different Mass Fractions of MON 810 Maize*, EC certification report EUR 20111 EN. EurOP, Luxembourg, Belgium.

Trapmann, S., Schimmel, H., Brodmann, P., Van den Eede, G., Vorburger, K., and Pauwels J. (2000). Influence of DNA degradation on the quantification of GMOs, poster at Joint Workshop on Method Development in Regulation to Regulatory Requirements for the Detection of GMOs in Food Chain, December 11-13, Brussels, Belgium.

Trapmann, S., Schimmel, H., Catalani, P., Conneely, P., Kramer, G.N., Le Guern, L., Prokisch, J., Robouch, P., Zeleny, R., and Pauwels J. (2001) Production of third generation RR-soybean reference material, poster at 115th AOAC International Annual Meeting and Exposition, September 9-13, Kansas City, Missouri.

Trapmann, S., Schimmel, H., Kramer, G.N., Van den Eede, G., and Pauwels, J. (2002). Production of certified reference materials for the detection of genetically modified organisms. *J. AOAC Int.* 85(3):775-779.

Trapmann, S., Schimmel, H., Le Guern, L., Zeleny, R., Prokisch, J., Robouch, P., Kramer, G.N., and Pauwels, J. (2000). Dry mixing techniques for the production of genetically modified maize reference materials, poster at Eighth International Symposium on Biological and Environmental Reference Materials (BERM-8), September 17-22, Bethesda, MD.

U.S. Department of Agriculture (USDA) (2002). *Sampling for the Detection of Biotech Grains.* <http://www.usda.gov/gipsa/biotech/sample2.htm>.

Van den Eede, G., Schimmel, H., Kay, S., and Anklam, E. (2002). Analytical challenges: Bridging the gap from regulation to enforcement. *J. AOAC Int.* 85(3): 757-761.

Chapter 5

Protein-Based Methods: Elucidation of the Principles

Farid E. Ahmed

INTRODUCTION

Immunoassays are ideal techniques for qualitative and quantitative detection of a variety of proteins in complex material when the target analyte is known (Ahmed, 2002). This chapter details the principles of immunoassay techniques widely used for genetically modified organism (GMO) detection, and the requirements for quality control, optimization, and validation standards.

The European Commission's (EC) Novel Food Regulation 258/7 and Council Regulation 1139/98 (EC, 1998) stipulate that novel foods and food ingredients that are considered to be "no longer equivalent" to conventional counterparts must be labeled. The presence of recombinant DNA, which produces modified protein in food products, is considered to classify these products as no longer equivalent if present at levels above 1 percent. The development and application of reliable and quantitative analytical detection methods are, therefore, essential for the implementation of labeling rules (Kuiper, 1999).

The ability to analyze a specific protein is not easy unless the protein has a unique property (such as that possessed by an enzyme) or a physical property (such as a spectrometric characteristic conferred upon by a nonprotein moiety). Nevertheless, it has been usually necessary to perform extensive, time-consuming, and specialized matrix

Thanks to Dean Layton of EnviroLogix, Inc., for the figures he provided (Figures. 5.1, 5.2, 5.6-5.8, and 5.10), and other colleagues who also permitted me to use their figures .

pretreatment prior to food analysis (Ahmed, 2001, 2003). Initially, bioassays based on whole animal responses were employed. In the late 1950s, however, immunoassays utilizing high affinity polyclonal antibodies resulted in high throughput, sensitive, and specific in vitro analytical procedures (Yallow and Benson, 1959). Subsequent developments of great significance in immunochemistry focused on two main areas: antibody production and assay formats (Harlow and Lane, 1999).

The immune system of higher animals has the ability to produce diverse responses to antigenetic stimuli, making itself capable of interacting with, and protecting itself against, the diversity of molecular and cellular threats. Each of the produced antibodies has a different structure, and each structure is produced by a different line or clone of cells. Antibodies are a large family of glycoproteins that share key structural and functional properties. Functionally, they can be characterized by their ability to bind both to antigens and to specialized cells or proteins of the immune system. Structurally, antibodies are composed of one or more copies of a characteristic unit that can be visualized as forming a Y shape (Figure 5.1). Each Y contains four polypeptides—two identical copies of a polypeptide known as the heavy chain and two identical copies of a polypeptide known as the light chain. Antibodies are divided into five classes—IgG, IgM, IgA, IgE, and IgD—on the basis of the number of the Y-like units and the type of heavy-chain polypeptide they contain (Ahmed, 1995). Many of the important structural features of antibodies can be explained by considering that IgG molecules having a molecular weight of ~160,000 kDa contain only one structural Y unit, and are abundantly available in the serum (MacCallum et al., 1996).

The antibodies-binding site on an antigen is known as an *epitope*, an area occupied by around 10 to 15 amino acids, some of which might be involved in low-affinity recognition around the fringes of the binding site (shown as a white hexagon in Figure 5.1). The epitope can be composed of amino acids sequential in the primary sequence to the protein. Such a sequence is known as a *continuous epitope*. Alternatively, the epitope might consist of amino acids distant in the primary sequence, but brought together by the force of secondary and tertiary structures. Such a recognition site is referred to as a *discontinuous epitope*. Disruption of secondary and tertiary structures can alter or abolish antibody recognition of a discontinuous

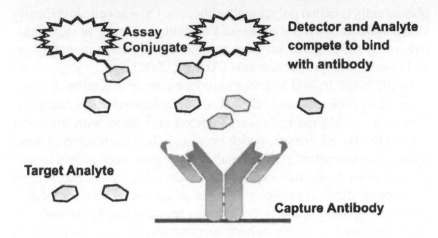

FIGURE 5.1. A competitive immunoassay showing an IgG molecule with four polypeptide chains and the binding sites (white hexagons), target analyte, and detectors competing to bind with antibodies (*Source:* Courtesy of Dean Layton, EnviroLogix, Inc.)

epitope. It is also possible that denaturation of a protein could also alter recognition of a continuous epitope, depending on the nature of the peptide (Wedemayer et al., 1997).

Most often, an animal will produce a large group of antibodies that recognizes independent epitopes on the antigen. Each type of antibody that recognizes a particular epitope is produced by a different clone of plasma cells, and each plasma cell can secrete antibodies that bind to only one epitope. In a given protein conformation, and with any assay format, antibodies are specific for the protein that do not bind to it, not because of the absence of the epitope, but because the epitope is hidden and not available for antibody recognition. Assay format has a profound effect on the availability of an epitope for binding by anditbody. The method of presentation of the protein to the antibody—even if mild—can also cause conformational change (Friguet et al., 1984) or can result in the hiding of an epitope, making antibody recognition and binding impossible for steric reasons (Brett et al., 1999). The collection of antibodies in serum that is synthesized by the concreted effort of a number of different antibody-producing plasma cells is called polyclonal antibodies. On a genetic level, polyclonal antibodies exhibit lot-to-lot variability, are more broadly reactive, often more sensitive, have shorter lead times to manufacture, and lower initial production cost (Ahmed, 2002).

In 1975, Köhler and Milstein introduced the first version of a procedure that produces antibodies of identical structure and reactivity. Precursors of plasma cells were isolated and fused with immortal cells of the B-cell lineage, which previously had been selected to no longer secrete antibodies. The resultant hybridoma cells could be single-cell cloned and then expanded as individual clones that secrete one type of identical antibody only, called monoclonal, with specific and easily studied properties of antigen recognition. These cells are commonly used as tissue culture supernatants that are harvested as spent media from growing the hybridomas. To produce large quantities, hybridoma cells are grown as tumors in syngenec mice, where they are secreted into an ascitic fluid that is then collected from the tumor-bearing mice with a high titer of monoclonal antibodies. The use of monoclonal antibodies enhanced the potential of two-site assays (Miles and Hales, 1968). In general, monoclonal antibodies show lot-to-lot consistency, can be supplied in indefinite quantities, are highly specific, require a long lead time to manufacture, and have a higher initial cost to produce (Ahmed, 2002).

Recent advances in antibody production, in which functional antigen-binding sites can be expressed on the surface of filamentous bacteriophages, make it feasible to produce antibodies in vitro, thereby extending the domain of potential antibody reactivity and permitting the manipulation of binding sites to improve specificity and affinity (McCafferty et al., 1990).

A number of factors affect the strength and specificity of the immune response, such as species of animal, physical state of the antigen, amount of antigen injected or introduced, route of injection, immunization schedule, and use of adjuvants (e.g., Freund's complete adjuvant) (Davies and Cohen, 1996). Antibodies can be purified by adsorption to and elution from beads coated with protein A, a component of the cell wall of the bacterium *Streptococcus aureus* (Hjelm et al., 1972).

PROBLEMS ASSOCIATED WITH DEVELOPING IMMUNOASSAYS FOR FOOD PROTEINS AND SUGGESTED REMEDIES

An understanding of how a protein behaves during food production and processing can help when attempting to use foods for antibody production, immunoassay assay development, and their use as an analytical standard (Beier and Stanker, 1996). Factors that aid in the development of sound immunologic assays for food proteins include: (1) whether the protein becomes modified post-translationally, and whether it is present in a homogenous format; (2) whether processing—including thermal and enzymatic treatments—causes peptides to fragment, and whether it is possible to utilize these fragments in the analytical procedure; (3) the prevalence of sequence homology between novel protein and other proteins normally present; and (4) the relationship between modified proteins and other proteins present in other food material. The presence of identical proteins in food may cause problems unless their sequence is not overly homologous to the target protein. In such a case, antibodies generated against the unique peptide sequence can be developed. Processing of food can also lead to extensive changes in protein conformation, making quantitation impossible (McNeal, 1988). This problem, however, can be overcome if antibodies directed against processing stable epitopes can be identified and produced (Huang et al., 1997).

The nature of antibodies, whether monoclonal or polyclonal, for detection of food protein is irrelevant. Polyclonal antibodies can be obtained in large amounts, are widely disseminated, and often exhibit better specificity. Nevertheless, the production of antibodies against particular peptides (rather than whole proteins) might be better achieved using monoclonal antibodies. Some immunoassay formats are better suited to monoclonal antibodies. For example, a two-site double-antibody sandwich (in which the antigen is localized between the capture antibody and the detection antibody) has superior properties to a competitive assay (in which the detector and analyte compete to bind with capture antibodies) (Pound, 1998), and works well with monoclonal antibodies. Moreover, some of the rapid immunoassays, such as the dipstick (or conventionally known as the *lateral flow strip*) assay, which consume large amounts of antibodies, may not be satisfied by polyclonal antibody preparations (Brett et al., 1999).

The following strategies and quality assurance (QA) issues are helpful to consider when setting up and validating an immunoassay for detection of food proteins.

Analytical Target

The uniqueness of the protein must be considered in both modified and unmodified foods. It may be necessary to identify, synthesize and use specific protein sequences as markers and standards. In addition, the target should not be the material used for antibody production (Brett et al., 1999).

Immunogen

The selection of the immunogen is critical. For polyclonal antibody production, extraneous material (e.g., nonspecific peptide sequences and impurities) should be eliminated from the preparations. For monoclonal antibodies, purity is less problematic because it is feasible to isolate the desired antibody during the selection process. Ideally, the screening process should be (1) identical to the final assay format; (2) would include ways of looking at matrix effects and assessments of sensitivity and specificity; and (3) be able to perform under extremes of temperature and time. If it is desirable to raise antibodies to a particular peptide sequence, selection of immunogens becomes problematic. This is because the antibodies recognize the peptide well, but fail to recognize the peptide sequence as present in the parent protein due to extensive structural mobility of the peptide compared to the more limited structures adopted when anchored within the protein. Moreover, complications might arise when the peptide is regarded as a hapten, and not as an antigen, in order to stimulate antibody production because the peptide will not conjugate with an immunogen or produce antibodies. For monoclonal antibody production, use of an intact protein as an immunogen, and use of a peptide to select the appropriate antibody may lead to a greater probability of selecting antibodies specific for a particular peptide sequence within the structure of the complete protein (Brett et al., 1999).

The rate-limiting steps in immunoassay development and application are (1) the need for widespread availability of appropriate antibodies, (2) the difficulties of producing antibodies to particular sequences from a protein, and (3) the inability to generate antibodies capable of reacting normally at extremes of temperature, pH, or high concentration of salt or solvent (Brett et al., 1999). The application of recombinant antibody technology may make it possible to more easily select antibodies with rare properties and to manipulate the properties of antibodies already available. As our understanding of these issues increase, it will become easier to grasp the changes undergone by polymers when subjected to denaturing conditions (Brett et al., 1999).

Immunoassay Format

Early considerations to ideas on final assay format such as the selection of immunogen and the type and properties of sought antibodies are important factors to keep in mind. Choice of format is determined by application, and a flexible format provides for diverse application. In a competitive assay, the unknown analyte and a marker analyte compete for a limited number of antibody binding sites (Figure 5.1). The amount of marker analyte that binds to antibody is determined by how much unknown is present. The unknown can be quantified by reference to the behavior of known standards (Ahmed, 2002).

On the other hand, in a sandwich assay (also known as two-site or reagent excess assay) an excess of the first antibody (referred to as a capture antibody)—which is normally immobilized to a solid surface—captures a sample of the analyte. An excess of a second antibody (usually labeled and known as a detector antibody) reacts with and labels the captured analyte (Figure 5.2). Quantitation is achieved by comparison with known standards. This assay is considered preferable to a competitive assay if (1) the analyte protein is large enough to allow binding of two antibodies at the same time (the sensitivity of the assay is not limited by the affinity of antibodies because it is a reagent excess assay), (2) the combination of different antibodies can increase specificity, and (3) the format is suited to a dipstick-type strip (Ekins, 1991).

The choice of the nature of the assay (i.e., quantitative versus qualitative) depends on its purpose. The ease and simplicity with which qualitative tests are applied is attractive because they do not require specialized laboratory or sophisticated equipment, and can be mastered with minimum training and skills. Immunoassays can also be

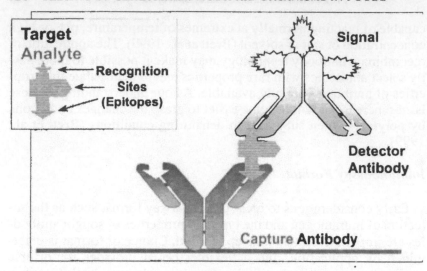

FIGURE 5.2. A sandwich (double-antibody) immunoassay showing a capture antibody and detector antibody (*Source:* Courtesy of Dean Layton, EnviroLogix, Inc.)

batch processed to produce quantitative and automatable results, which takes longer than qualitative tests, although the average time per sample might be similar. However, the information obtained from these quantitative tests outweighs the increased time, labor, and cost involved (Kuiper, 1999). Extraneous material that causes contamination problems in immunoassays comes largely from four main sources: plant sap and tissue, fungal and bacterial contamination of solutions and analytical reagents, dirt or dust from the environment, and workers' body fluids (i.e., saliva and sweat) (Sutula, 1996).

Assay Validation

Assay validation is the process of demonstrating that the combined procedures of sample pretreatment, preparation, extraction, cleanup, and analysis will yield acceptable, accurate, precise, and reproducible results for a given analyte in a specified matrix (Ahmed, 2001, 2003; Mihaliak and Berberich, 1995). Analysis of food is complicated by the variety of available matrixes, which put a strain on the analysis and the validation of results. Immunoassays have the advantage that large number of samples can be processed automatically in a short time making for easier assay validation, although there is no substitute for the experience of the operator and the availability of historical-control data (Ahmed, 2002; Brett et al., 1999). Establishing guidelines for the validation and use of immunoassays will help provide users with a consistent approach to adopting the technology, and will help produce accurate and meaningful results from the assays (Lipton et al., 2000; Sutula, 1996).

Confirmatory Method

Since no two antibody preparations are identical in their behavior, and since no two analytes will react in the same way with an antibody, then it is possible that by using different quantitative immunoassays—based on different antibodies—meaningful data can be generated to quantify cross-reacting material if present in the sample (Karu et al., 1994) or to confirm the correctness of obtained results. The importance of availability of positive and negative controls for immunoassays cannot be overemphasized (Lipton et al., 2000; Sutula, 1996).

Assay Sensitivity

The strength of binding of antigens to antibodies determines the sensitivity of the method. The lower limit of detection (LOD) or sensitivity, for a number of routine clinical immunoassays is in the order of 10^{-12} to 10^{-13} (Khalil, 1991).

Limit of quantitation (LQ) is the smallest concentration of analyte that can be measured in samples and yield predicted concentrations with an acceptable level of precision and accuracy (Rittenburg and Dautlick, 1995). LQ is commonly defined as a concentration equal to the lowest standard used in the assay. On the other hand, *quantitative range* (QR) is the lower and upper limits of analyte concentration over which the method yields quantitative results within the stated performance limit. QR is determined by the range of analyte concentration used to construct the standard curve (Lipton et al., 2000).

Figure 5.3a, which shows a typical dose-response curve for a transgenic protein using a standard sandwich ELISA method, reveals that the LOD is approximately 10^{-12}, in accordance with other established protein immunoassays. The response of a sandwich ELISA

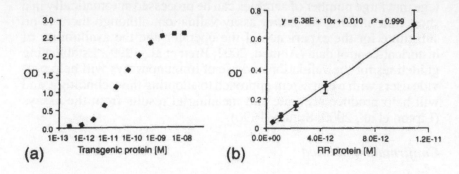

FIGURE 5.3. Dose-response curves for ELISA assay. (a) Typical response curve for a purified transgenic protein in a sandwich ELISA. Assuming the protein has a molecular weight of 50,000, a minimum detection limit of 10^{-12} translates to 50 pg/ml. (b) Detection of the purified, novel protein found in RR crops by a commercial sandwich ELISA assay. The equation represents the linear regression estimate of the line defined by the data points. The correlation coefficient is given by r^2. Error bars represent ±2 SD (95 percent confidence) around the means ($n = 3$). (*Source:* Reprinted from *Food Control*, Vol. 10, Stave, J. W., Detection of new or modified proteins in normal foods derived from GMO—future needs, pp. 367-374. Copyright 2001, with permission from Elsevier.)

format to purified Roundup Ready (RR) protein at the lower limit of detection of the method can be seen in Figure 5.3b. Experience with immunoassays for protein detection has shown that the concentration of normal protein that must be detected is 6×10^{-10} M, which is well within the detection limit of protein immunoassays. Moreover, immunoassays employing higher sensitivity detection methods (e.g., chemiluminescence) were reported to detect protein concentrations in the order of 10^{-12} (Khalil, 1991). It has been suggested that very low concentrations of GMOs may be detected in raw agricultural commodities (e.g., seed, grain) by testing greater numbers of individual seeds (Stave, 1999).

Assay Specificity

The degree to which analogs or other molecules bind to antibodies, i.e., cross-reactivity, is a measure of specificity that should be characterized and detailed in the method. Potential for interferences from reagents and laboratory ware can be evaluated by assaying extracts from nontransgenic plant material. If a substance in the final extract —other than the specific protein analyte—affects the response of the method, the nonspecific response is referred to as a *matrix effect* (Ahmed 2001, 2003). One way to manage matrix effects is to demonstrate that the analytical method employed in all matrixes gives identical results with or without the matrix present in the extract. Another way is to prepare the standard solutions in extracts from a nontransgenic matrix (i.e., matrix-matched standards). This would ensure that any matrix effect is consistent between the standards and the sample (Lipton et al., 2000; Stave, 1999).

Extraction Efficiency

Extraction efficiency is a measure of how efficient a given extraction method is at separating the protein analyte from the matrix, and is expressed as percent analyte recovered from the sample. Because the expressed protein is endogenous to plants, it may be difficult to demonstrate efficiency of the extraction procedure, as there may not be an alternate detection method against which to compare results. An approach to addressing this difficulty is to demonstrate the recovery of each type of introduced protein analyte from each type of food fraction by exhaustive extraction, i.e., repeatedly extracting the sample until no protein is detected (Stave, 1999).

Precision (Accuracy) of Immunoassays

Intra-assay precision is a measure of variations (or degree of randomness) occurring within the assay. It can be evaluated by determining standard deviation (SD), or percent coefficient of variation (percent CV) between replicates assayed at various concentrations on the standard curve, and on the pooled variations—derived from absorbance values in standards from independent assays—performed on different days. The greater the precision, the closer two concentrations can be and still be differentiated by the assay (Ekins, 1991). On the other hand, interassay precision describes how much variation occurs between separate assays, and is measured by analysis of quality control (QC) samples consisting of two pooled extracts, one from transgenic plant tissue and another from nontransgenic tissue, in every assay (Mihaliak and Berberich, 1995). Interassay precision can be evaluated over time and is expressed as percent CV (Rogan et al., 1992, 1999).

The accuracy of an immunoassay is considered acceptable if the measured concentration is between 70 and 120 percent of the actual concentration, and a CV of less than 20 percent for measured recoveries is achievable at each fortification level (Poulsen and Bjerrum, 1991).

Assay Reproducibility

This is a measure of obtaining consistent results from test to test, or from analyst to analyst, on assays carried out from one day to another. It can be determined by measuring variations among replicates within the test, among analysts, and among separate tests from day to day (Ahmed, 2002).

Ruggedness

This measures how well the assay performs under various conditions, including use of second party validation. Assays carried out to establish ruggedness include repeated analysis of a sample or samples on several days, and measuring accuracy and precision in fortified samples employing controls derived from various sources (Ahmed, 2002; Lipton et al., 2000).

Availability of Testing Material and Standards

Until now, most of the novel proteins that have been introduced into transgenic crops are proprietary to the company developing the GMO, which limits their availability to outside parties. To enable objective testing, companies developing GMOs containing novel proteins must either provide validated testing methods, or make antigens, antibodies, or the means to produce them accessible to outsiders interested in analytical test development, in addition to providing standardized reference material to these proteins and depositing them with an agency suited for distributing them at a minimum charge (Ahmed, 2002; Stave, 1999).

During assay development, the reference material would be used to help define assay parameters that would minimize any interfering matrix effect, such as nonspecific binding of sample components to antibodies. During validation and use of the assay, the reference material would be treated, extracted, and analyzed alongside the test samples in order to compare results (Ahmed, 2002; Lipton et al., 2000).

Because it is unlikely that reference material will be developed for every modified food, a practical solution may lie in developing the reference standards for major commodities and ingredients, and instituting tests at specific critical control points (CCPs) in the food cycle production (Ahmed, 1992). It has been suggested that downstream from a CCP, the concentration of GMO in the food could be estimated from the process characteristics and verified through documentation of chain of custody (Stave, 1999).

Use of Instrumental Techniques

Advances are being made in combining antibodies with instruments, such as hyphenated methods employing multidimensional liquid chromatography and tandem mass spectrometry (Link et al., 1999), and nuclear magnetic resonance technology (Williams, 2002). Advances have been made in real-time detection employing antibodies immobilized onto the surface of physicochemical transducers, giving rise to biosensors, or immunosensors, when antibodies are the biorecognition system (Brett et al., 1999; Kuiper, 1999).

WESTERN BLOTTING

Western blotting is a highly specific method that provides qualitative results suitable for determining whether a sample contains a target protein below or above a predetermined threshold value, and may also provide quantitative results when using appropriate standards (Ahmed, 2002). It employs electrophoretic separation of denatured proteins from a gel to a solid support, and probes proteins with reagents specific for particular sequences of amino acids, followed by bound antibody detection with one of several secondary immunological reagents, i.e., ^{125}I-labeled protein A or anti-immunoglobulin, or anti-immunoglobulin on protein A coupled to horseradish peroxidase (HRP) or alkaline phosphatase (AP), which converts a substrate into a colored, fluorescent, or chemiluminescent product (Burnette, 1981). The proteins are usually antibodies specific for antigenic epitopes displayed by target proteins. The technique is sensitive and specific for protein detection in complex mixtures of proteins that are not radioactive. Moreover, since electrophoretic separation of protein is mostly carried out under denaturing conditions, problems due to solubilization, aggregation, or coprecipitation of target protein with adventitious proteins are usually avoided. Not all monoclonal antibodies are suitable as probes when the target proteins are highly denatured, and reliance on a single antibody is unwise because of the high frequency of spurious cross-reactions with irrelevant proteins. On the other hand, when employing polyclonal antibodies, the efficiency with which a given polyclonal antibody can detect different antigenic epitopes of immobilized and denatured target proteins is often unpredictable. As little as 1 ng of protein can be detected by Western blotting (Sambrook and Russel, 2000).

The anionic detergent sodium dodecyl sulfate (SDS) is used together with a reducing agent and heat to dissociate proteins before they are loaded onto the gel. The denatured polypeptides complex with SDS and become negatively charged. Because the bound SDS is proportional to the molecular weight of the polypeptide and is independent of its sequence, the complex migrates toward the anode through the gel according to the size of the polypeptide. SDS-polyacrylamide gel (PAG) electrophoresis is carried out utilizing a discontinuous buffer system in which the buffer in the reservoirs of the apparatus is of different pH and ionic strength from the buffer used to cast the gel (Laemmli, 1970). When an electric current is applied, the sample loaded to the well of the gel is swept along a moving boundary, and the complex is deposited in a thin zone on the surface of the resolving, stacking gel. The chloride ions in the sample and stacking gel form the leading edge of the moving boundary, and the trailing edge contains glycine molecules. In between is a zone of lower conductivity and steeper voltage gradient that sweeps the polypeptides from the sample, depositing them on the surface of the gel, where the higher pH favors ionization of glycine. Glycine ions migrate through the stacked peptides and travel behind chloride ions. Freed from the moving boundary, the SDS-PAG complex moves through the gel in a zone of uniform voltage and pH, and separates according to size by sieving (Sambrook and Russel, 2000).

If the Western blot is to be probed with a radiolabeled antibody or radiolabeled protein A, the immobilized protein may be stained with India ink, which is cheaper and more sensitive than Ponceau S, and provides a permanent record of the location of proteins on the nitrocellulose or polyvinylidene difluoride (PVDF) membranes. Staining with colloidal gold for total protein detection followed by immunostaining of individual antigens has been reported (Egger and Bienz, 1998).

The sensitivity of Western blotting depends on reducing nonspecific proteins present in the immunological reagents used for probing, which can also transfer from SDS-PAG. The least expensive convenient blocking solution is nonfat dried milk (at 5 percent), provided probed proteins are different from those present in milk (Johnson et al., 1984).

The effective range of separation of gels depends on the concentration of the polyacrylamide used to cast the gel and the amount of cross-linking produced when adding a bifunctional agent such as N, N'-methylenebisacrylamide, which adds rigidity to the gel and forms pores through which the SDS-polypeptide complex passes. The size of these pores decreases as the bisacrylamide:acrylamide ratio increases, reaching a minimum when the molar ratio is about 1:29. This has been shown empirically to resolve peptides that differ in size by as little as 3 percent (Harlow and Lane, 1999). The linear range of separation obtained with gel casts containing an acrylamide concentration ranging from 5 to 15 percent has been from 200 to 12 kDa, respectively (Ornstein, 1964). SDS-PAG can be stained with Coomassie Brilliant Blue A250, silver salts (e.g., $AgNO_3$), then dried and photographed. Immobilized proteins can be labeled with alkaline phosphatase. The location of such an enzyme can be revealed by treatment with a chromogenic substrate containing 5-bromo-4-chloro-3-indolylphosphate (BCIP) and nitroblue tetrazolium (NBT), which produces a purple, insoluble end product (Pluzek and Ramlau, 1988). Chemiluminescent reagents catalyzed by horseradish peroxidase (e.g., ECL by Amersham Biosciences) in combination with a fluorescent scanner have also been used (Thorpe and Kricka, 1986). Visualization of protein blots is depicted diagrammatically in Figure 5.4.

Western blotting is less susceptible to matrix effects and protein denaturation when used in analysis of food proteins (Ahmed, 2002;

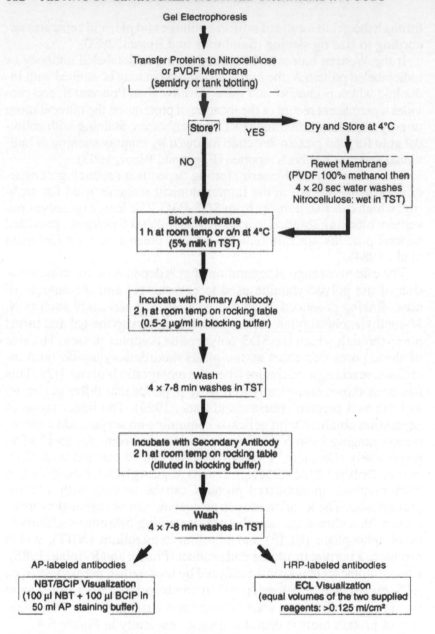

FIGURE 5.4. Flow diagram for visualization of protein blots (*Source:* From Finney, 1998. Reprinted with permission from Humana Press.)

Rogan et al., 1999). Monoclonal antibodies generated against the enzyme 5-enolpyruvylshikimate-3-phosphate synthase (EPSPS) derived from the *Agrobacterium tumefaciens* strain CP4, which provides the glyphosate tolerance to Roundup herbicide in Roundup Ready (RR) soy plants, were prepared by hybridoma technology and purified by immunoaffinity chromatography. This enabled the detection of 47 kDa CP4 synthase in RR soy at a sensitivity of between 0.5 and 1 percent in raw soy and soy protein fractions, but not in processed soy ingredients (van Duijn et al., 1999).

Another Western blot assay utilized a polyclonal goat anti-CP4 ESPSP antibody, followed by detection with biotinylated protein G using HRP-labeled NeutrAvidin and a signal developed by chemiluminescence. Results from validation studies showed that a Laemmli or Tris extraction buffer supplemented with a zwitterionic (i.e., neutral) detergent, such as CHAPS, and a chaotropic salt such as guanidine hydrochloride, provided the most efficient extraction buffer for highly processed soy fractions (Figure 5.5), with toasted meal showing the lowest CP4 EPSPS content (30 percent of the level observed for unprocessed seed). However, levels of protein concentrate and protein isolate were not reduced from that observed for unprocessed soybean seed. The accuracy and precision of Western blotting was evaluated using a blind comparison test, showing that "trained" personnel can correctly and consistently identify samples containing around 2 percent soybean. The detection limit (sensitivity) was 100 pg of CP4 EPSPS protein, and ranged from 0.25 percent (for seed) to 1 percent (for toasted meal) for various soybean fractions. Western blotting would be most effectively utilized to determine whether a sample contained CP4 EPSPS protein equal to or above a predetermined level (i.e., threshold detection) (Rogan et al., 1999).

This technique—although sensitive and able to produce quantitative results—is used mostly as a research tool for large-scale testing of GMOs because it is not amenable to automation. Western blotting generally takes about two days and costs about $150/sample.

IMMUNOASSAYS

An *immunoassay* is an analytical procedure that employs the specific binding of an antibody with its target protein analyte. Anti-

1 2   3 4 5 6 7 8   9 10 11 12 13 14 15   1 2   3 4 5 6 7 8 9 10 11 12 13 14 15

Soybean Seed   Toasted Meal

Gel Loading Information	Buffer Information
Lane Description 1 Transfer markers 2 CP4 EPSPS std., 5 ng 3 Control,* buffer 1, 5 µl 4 Control, buffer 2, 5 µl 5 Control, buffer 3, 5 µl 6 Control, buffer 4, 5 µl 7 Control, buffer 5, 5 µl 8 Control, buffer 6, 5 µl 9 Transgenic*, buffer 1, 5 µl 10 Transgenic, buffer 2, 5 µl 11 Transgenic, buffer 3, 5 µl 12 Transgenic, buffer 4, 5 µl 13 Transgenic, buffer 5, 5 µl 14 Transgenic, buffer 6, 5 µl 15 Transfer markers * Refers to either soybean seed or toasted meal extracts.	**# 1**: 100 mM Tris-Cl, pH 7.5, 1 mM benzamidine-HCl, 5 mM DTT, 2.5 mM EDTA, 1.0 mM PMSF. **# 2**: # 1 with 150 mM Kcl **# 3**: # 1 with 150 mM KCl and 10 mM CHAPS **# 4***: # 1 with 10 mM CHAPS and 6 M guanidine-HCl **# 5**: 1X Laemmli buffer [62.5 mM Tris-Cl, pH 6.8, 2% (w/v) SDS, 10% (v/v) glycerol, 0.05% (w/v) bromophenol blue] **# 6**: Buffer 1 with 1% (v/v) Tween-20 All samples were extracted at a ratio of 80 mg tissue to 4.0 ml of buffer and prepared as 1X Laemmli samples before analysis. * Extracts were desalted into buffer #1 before further analysis.

FIGURE 5.5. Effect of different buffers on the extraction of CP4 EPSPS protein from RR soybean seeds and toasted meal samples. Data show that toasted meal requires anionic or chaotropic salts (e.g., guanidine-HCl) in combination with a neutral detergent such as CHAPS. Lanes 1 and 15 were loaded with a color marker (Amersham Biosciences RPN 756). Lane 2 was loaded with 5 ng of a standard CP EPSPS produced in the bacterium *Escherichia coli.* Lanes 3, 4, 5, 6, 7, and 8 represent control samples, whereas lanes 9, 10, 11, 12, 13, and 14 represent transgenic samples. Both were extracted in buffers 1, 2, 3, 4, 5, and 6, respectively. Buffer 1 was 100 mM Tris-Cl, 1 mM benzamidine-HCl, 5 mM DTT, 2.5 mM EDTA, and 1.0 mM PMSF. Buffer 2 was buffer 1 with 150 mM KCl. Buffer 3 was buffer 1 with 150 mM KCl and 10 mM CHAPS. Buffer 4 was buffer 1 with 10 mM CHAPS and 6 M guanidine-HCL. Buffer 5 was 1X Laemmli buffer: 62.5 mM Tris-Cl, pH 6.8, 2% (w/v) SDS, 10% (v/v) glycerol, 5% (v/v) 2-mercaptoethanol, and 0.05% (v/v) bromophenol blue. Buffer 6 was buffer 1 with 1% (v/v) Tween-20. All samples were extracted at a ratio of 80 mg tissue to 4.0 mL of buffer and prepared as 1X Laemmli samples before analysis. (*Source:* Reprinted from *Food Control,* Vol 10, Rogan et al., Immunodiagnostic methods for detection of 5-enolpyruvylshikimate-3-phosphate synthase in Roundup Ready soybeans, pp. 407-414, copyright 1999, with permission from Elsevier.)

bodies are commonly attached to a solid phase, such as plastic wells, tubes, capillaries, membranes, and latex and magnetic particles. Detection is accomplished by adding a signal-generating component (reporter antibody) that responds to the target protein. Various labels (chromagens) have been employed, including radioactivity, enzymes, fluorescence, phosphorescence, chemiluminescence, and bioluminescence (Gee et al., 1994), which interact with the substrate, leading to detection (Figure 5.6).

At present, microwell plate and coated tube sandwich ELISAs (Figures 5.7 and 5.8, respectively) and lateral flow strip (Figure 5.9) assays are the most common formats employed for GMO detection. Such assays use enzymes and colorimetric substrates to produce a colored product. The microwell plate ELISA is used as a qualitative (yes/no) assay, a semiquantitative (threshold) assay, or a quantitative assay, whereas lateral flow devices give a yes/no result or an indication of whether the sample contains the target analyte at or above an

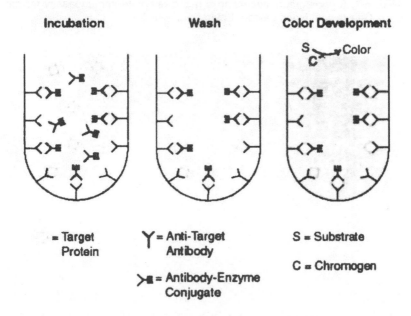

FIGURE 5.6. Principles of immunoassays showing sequential incubation, washing, and color development (detection) steps (*Source:* Courtesy of Dean Layton, EnviroLogix, Inc.)

FIGURE 5.7. A photograph illustrating a microwell plate immunoassay format (*Source:* Courtesy of Dean Layton, EnviroLogix, Inc.)

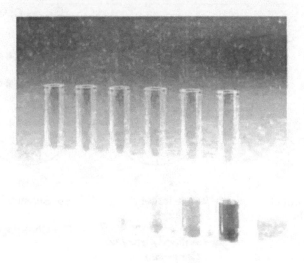

FIGURE 5.8. A photograph illustrating the coated tube immunoassay format. (*Source:* Courtesy of Dean Layton, EnviroLogix, Inc.)

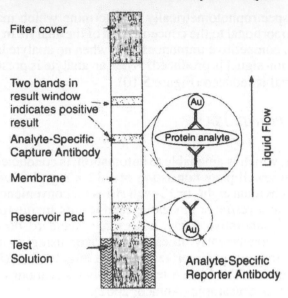

FIGURE 5.9. A schematic view of the lateral flow strip format illustrating the principles of the assay (*Source:* Reprinted from *Food and Agriculture Immunology,* Vol. 12, Lipton et al., Guidelines for the validation and use of immunoassays for determination of introduced proteins in biotechnology enhanced crops and dreived food additives, pp. 153-164, copyright 2000, with permission from Taylor & Francis, Ltd., <http://www.tandf.co.uk/journals>.)

established threshold level. The choice of which immunoassay format to use is usually determined by the intended application (Ahmed, 2002; Lipton et al., 2000).

Sandwich immunoassays that involve immobilization of a capture antibody on a solid support, for example, plastic walls of a microwell plate or on membranes (Figure 5.2), are currently the most commonly employed format. When an analyte and enzyme-labeled reporter antibodies are added, an antibody-analyte binding occurs. Following incubation, the unbound analyte and reporter antibodies are washed away. A second analyte-specific-labeled antibody (detection antibody) is added, which also binds to the analyte, forming a sandwich. When a colorimetric substrate is added, it reacts with the enzyme forming a colored product that can be observed visually or measured spectrophotometrically to determine which magnitude is directly proportional to the concentration of the analyte in the sample solution. In competitive immunoassays, when no analyte is present, a high detection signal is produced. When an analyte is present, the detection signal is reduced (Figure 5.10).

Microwell Plate ELISA

This test, which is amenable to automation, is conducted in a standard 96-microwell plate consisting of a 12 × 8 grid of well (Figure 5.7) or as individual eight- or 12-well strips for convenience. This test can be used as a yes/no or threshold test if results are interpreted visually. For a quantitative assay, however, a standard dose-response curve is run concurrently on each microplate using an appropriate standard of the transgenic protein. Coated tube formats (Figure 5.8) are also available. The ELISA test generally takes about 1 to 2 hours, and costs about $5/sample (Ahmed, 2002).

I. No analyte—high detection signal

II. Analyte present—detection signal reduced

FIGURE 5.10. Principles of a competitive immunoassay. (I) A high detection signal is produced when no analytes are present. (II) Detection signal is reduced in the presence of an analyte that competes for the binding sites. (*Source:* Courtesy of Dean Layton, EnviroLogix, Inc.)

Dot Blots and Lateral Flow Strips

Nitrocellulose and nylon membranes can also be used as the solid phase instead of microtiter plates. Such "dot blot" assays are useful for a small number of samples and they obviate the need for the initial serological trapping of antigens. A precipitating, rather than a soluble substrate, is used in the final deployment to visualize enzyme activity (De Boer et al., 1996).

A lateral flow strip is a single unit device that allows for manual testing of individual samples. Each nitrocellulose strip consists of three components: a reservoir pad on which an antibody coupled to a colored particle, such as colloidal gold or latex, is deposited, a result window, and a filter cover (Figure 5.9). An analyte-specific capture antibody is also immobilized on the strip. The test is conducted by inserting the strip in a tube containing the test solution. The solution moves toward the reservoir pad, solubilizing the reporter antibody, which binds to the target analyte and forms an analyte-antibody complex that flows with the liquid sample laterally along the surface of the strip. When the complex passes over the zone where the capture antibody has been immobilized, it binds to the antibody and produces a colored band. The presence of two bands indicates a positive test for the protein of interest. A single band indicates that the test was performed correctly, but that there was no protein of interest present in the sample. This test, which provides a yes/no or a threshold (semiquantitative) determination of the target protein in about 10 minutes at an average cost of $2/sample, is appropriate for field or on-site applications, and does not require specialized equipment (Ahmed, 2002).

Other Immunoassay Formats

In addition to the formats mentioned previously, other immunoassays utilize particles as the solid support. These particles are coated with the capture antibody and the reaction takes place in the test tube. Particles with bound reactants are separated from the unbound reactants in solution by a magnet. Advantages of this format include better reaction kinetics, because the particles move freely in the solution, and increased precision due to uniformity of the particles. Other, less commonly used formats utilize different enzyme tags and substrates, employ competitive assays, and utilize nesting or a combination of two steps in one (Ahmed, 2002; Brett et al., 1999; Lipton et al., 2000; Rogan et al., 1999; Stave, 1999).

Validation Studies

Several studies to validate ELISA and lateral flow strips have been carried out. A double antibody sandwich ELISA, in which antibodies are immobilized on microtiter plates, was developed, optimized, and validated to accurately quantitate neomycin phosphotransferase II (NPTII) levels in genetically modified cottonseed and leaf tissue. Using this assay, antibodies were produced in two New Zealand white rabbits following injection with NPTII in Freund's complete adjuvant. The IgG fractions from rabbit sera were affinity purified on protein-A-Sepharose CL4-B columns, followed by conjugation of antibodies to HRP. Validation of ELISA for NPTII expression in cottonseed and leaf extracts was carried out using Western blotting, and a functional assay (McDonnell et al., 1987) showed that the plant expressed protein. The NPTII standards were indistinguishable. The amount of NPTII present was directly proportional to the amount of peroxidase-labeled antibodies bound in the sandwich. Data on sensitivity, accuracy, and precision substantiated the utility of this ELISA for the analysis of NPTII in genetically modified plant tissue. Typical intra-assay variability (percent CV) was less than 5 percent for concentrations of NPTII (between 0.1 and 6 ng/well). Expression levels of NPTII measured in three different genetically modified cotton lines ranged from 0.040 to 0.044 percent and from 0.009 to 0.019 of extractable protein from leaf and seed extracts, respectively (Rogan et al., 1992).

An ELISA for measurement of CP4 EPSPS in a triple antibody sandwich procedure utilizing a monoclonal capture antibody and a polyclonal detection antibody followed by a third, biotin-labeled monoclonal antirabbit antibody using HRP-labeled NeutraAvidin for color visualization showed detection from 0.5 to 1.4 percent RR soybean for the various processed soybean fractions. Mean CP4 EPSPS expression was 210 µg/g tissue fresh weight and ranged from 195 to 220 µg/g tissue fresh weight, resulting in an overall variation of less than 15 percent CV (Rogan et al., 1999).

Another blind study on RR soybeans carried out in 38 laboratories in 13 countries was validated on matrix-matched reference material in the form of dried, powdered soybeans. The method employed a sandwich ELISA with monoclonal antibodies raised against the protein CP4 EPSPS, immobilized in the wells of ELISA plate, and a polyclonal antibody conjugated with HRP as the detection system using 3-3', 5, 5' tetramethylbenzidine (TMB) as the substrate for HRP detection. Detection limit was in the range of 0.35 percent GMO on a dry weight basis. Semiquantitative results showed that a given sample containing less than 2 percent genetically modified (GM) soybeans was identified as below 2 percent with a confidence of 99 percent. Quantitative results showed a repeatability (r) and reproducibility (R) to be $RSD_r = 7$ percent and $RSD_R = 10$ percent, respectively, for samples containing 2 percent GM soybeans (Lipp et al., 2000).

A study assessing the field performance of lateral flow strip kits to detect RR GMOs was carried out in 23 grain-handling facilities in the Midwest region of the United States (e.g., Iowa, Illinois, Indiana, Minnesota, and Ohio) to analyze a series of blinded samples containing defined proportions of conventional and transgenic soybeans. The observed rate of false positives was 6.7 percent when 1 percent GMOs was used, and 23.3 percent when 10 percent GMOs was used. The frequencies of false negatives were 93.3, 70.5, 33.3, 31.8, and 100 percent for samples containing 0.01, 0.1, 0.5, 1, and 10 percent GMOs. Based on these results, the authors interpreted that the kit—as a field tool—is not effective in monitoring GMOs at the level of 1 percent or lower. Moreover, statistical analysis of their results appeared to support the notion that limitations on operators' performance and not defects in kit material were the primary cause for the erroneous results, whereas sample size was suggested by the authors to play a secondary role (Fagan et al., 2001).

CONCLUSIONS

Many countries (e.g., those in the EU) are required to employ protein-based methods together with DNA-based methods for GMOs detection. Although protein-based tests are practical and cost-effective, some GM products do not express detectable levels of protein. Western blotting appears to be a robust procedure for detecting proteins that are easily degraded during processing. It offers both qualitative and semiquantitative estimates, and can detect proteins at 100 pg (or

~0.25 percent GMOs), although it is used mostly for research purposes instead of routine testing because it not easily amenable to automation (Ahmed, 2002).

ELISA methods appear to be the methods of choice for screening new materials and basic ingredients when the presence of modified proteins can be detected and is not degraded by processes such as extreme heat, pH, salts, etc. These assays are fast, robust, suitable for routine testing, and relatively inexpensive, with detection limits for GMOs in the order of 0.5 to 1 percent. However, their applicability is limited and dependent on the degree and nature of processing that the food has undergone (Kuiper, 1999). Food processing degrades proteins, and a negative list of ingredients—based on extensive knowledge of processing methods and of the composition of food ingredients—should be established (e.g., highly refined oil or processed starch products) in order to focus on detection of more relevant food ingredients. A need exists to develop new multidetection methods that are flexible, adjustable, and able to detect novel proteins when information on modified gene sequence are not available. Successful implementation of testing will be influenced by factors such as cost per test, ease of use, truncation time, and quality control considerations (Ahmed, 2002; Kuiper, 1999; Lipton et al., 2000; Stave, 1999).

Efforts must be accelerated to develop standardized reference material for these assays (Ahmed, 2002; Stave, 1999). False positive results should be avoided because they may result in destruction or unnecessary labeling of foods in countries that have labeling requirements, such as the EU. Established guidelines for validation and use of these immunoassays will provide users with a consistent approach to adopting the technology, and will help them produce accurate testing results (Lipton et al., 2000). Proprietary proteins must be made available by the producing companies to those interested in test development, and standard reference materials must be developed to ensure standardized test performance (Ahmed, 2002; Stave, 1999).

REFERENCES

Ahmed, F.E. (1992). Hazard analysis critical control point (HACC) as a public health system for improving food safety. *Int. J. Occup. Med. Toxicol.* 1:349-359.
Ahmed, F.E. (1995). Transgenic technology in biomedicine and toxicology. *Env. Tox. Ecotox. Revs.* C13:107-142.

Ahmed, F.E. (2001). Analyses of pesticides and their metabolites in foods and drinks. *Trends Anal. Chem.* 22:649-655.

Ahmed, F.E. (2002). Detection of genetically modified organisms in foods. *Tr. Biotechnol.* 20:215-223.

Ahmed, F.E. (2003). Analysis of polychlorinated biphenyls in food products. *Trends Anal. Chem.* 22:344-358.

Beier, R.C. and Stanker, L.H. (1996). *Immunoassays for Residue Analysis: Food Safety*. American Chemical Society, Washington, DC.

Brett, G.M., Chambers, S.J., Huang, L., and Morgan, M.R.A. (1999). Design and development of immunoassays for detection of proteins. *Food Control* 10:401-406.

Burnette, W.N. (1981). Western blotting: Electrophoretic transfer of proteins from sodium dodecyl sulfate-polyacrylamide gels to unmodified nitrocellulose and radiographic detection with antibody and radioiodinated protein A. *Anal. Biochem.* 112:95-112.

Davies, D.R. and Cohen, G.H. (1996). Interaction of protein antigens with antibodies. *Proc. Natl. Acad. Sci. USA* 93:7-12.

De Boer, S.H., Cuppels, D.A. and Gitaitis, R.D. (1996). Detecting latent bacterial infections. *Adv. Botan. Res.* 23:27-57.

Egger, D. and Bienz, K. (1998). Colloidal gold staining and immunoprobing on the use of Western blot. In *Immunochemical Protocols*, Second Edition (pp. 217-223), J.D. Pounds, ed. Humana Press, Totowa, NJ.

Ekins, R.P. (1991). Competitive, noncompetitive, and multianalyte microspot immunoassay. In *Immunochemistry of Solid-Phase Immunoassay* (pp. 215-223), J.E. Butler, ed. CRC Press, Boca Raton, FL.

European Commission (EC) (1998). Council Regulation No 1139/98 of May 26, concerning the compulsory indication of the labeling of certain foodstuffs produced from genetically modified organisms of particulars other than those provided for in Directive 79/1, 12. *EEC Official J. L.* 159(03106/1998):4-7.

Fagan, J., Schoel, B., Huegert, A., Moore, J., and Beeby, J. (2001). Performance assessment under field conditions of a rapid immunological test for transgenic soybeans. *Int. J. Food Sci. Technol.* 36:1-11.

Finney, M. (1998). Nonradioactive methods for visualization of protein blots. In *Immunochemical Protocols*, Second Edition (pp. 207-216), J.D. Pounds, ed. Humana Press, Totowa, NJ.

Friguet, B., Djavadi-Ohaniance, L., and Goldberg, M.E. (1984). Some monoclonal antibodies raised with a native protein bind preferentially to the denatured antigen. *Molec. Immunol.* 21:673-677.

Gee, S.J., Hammock, B.D., and Van Emon, J.M. (1994). *A User's Guide to Environmental Immunochemical Analysis*. U.S. EPA Office of Research and Development, Washington, DC.

Harlow, E. and Lane, D. (1999). *Using Antibodies: A Laboratory Manual*. Cold Spring Harbor Laboratory Press, Cold Spring Harbor, NY.

Hjelm, H., Hjelm, K., and Sjöquist, J. (1972). Protein A from *Staphylococcus aureus*. Its isolation by affinity chromatography and its use as an immunosorbent for isolation of immunoglobulins. *FEBS Lett.* 28:73-85.

Huang, L., Brett, G.M., Mills, E.N.C., and Morgan, M.R.A. (1997). Monoclonal antibodies as molecular probes for thermal denaturation of soya protein. In *Proceedings of Antibodies in Agrifood Science: From Research to Application* (pp. 17-24). Hanover, New Hampshire.

Huang, L., Mills, E.N.C., Carter, J.M., Plumb, G.W., and Morgan, M.R.A. (1998). Analysis of thermal stability of soya globulins using monoclonal antibodies. *Biochim. Biophy. Acta* 1388:215-226.

Johnson, D.A., Gautsch, J.W., Sportsman, J.R., and Elder, J.H. (1984). Improved technique utilizing nonfat dry milk for analysis of proteins and nucleic acid transferred to nitrocellulose. *Gene Anal. Tech.* 1:3-12.

Karu, A.E., Lin, T.H., Breiman, L., Muldoon, M.T., and Hsu, J. (1994). Use of multivariate statistical methods to identify immunochemical cross-reactants. *Food Agric. Immunol.* 6:371-384.

Khalil, O.S. (1991). Photophysics of heterogeneous immunoassays. In *Immunochemistry of Solid-Phase Immunoassay* (pp. 113-121), J.E. Butler, ed. CRC Press, Boca Raton, FL.

Köhler, G. and Milstein, C. (1975). Continuous culture of fused cells secreting antibody of predefined specificity. *Nature* 256:495-497.

Kuiper, H.A. (1999). Summary report of the ILSI Europe workshop on detection methods for novel foods derived from genetically modified organisms. *Food Control* 19:339-349.

Laemmli, U.K. (1970). Cleavage of structural proteins during the assembly of the head of bacteriophage T4. *Nature* 227:680-695.

Link, A.J., Eng, J., Schieltz, D.V., Carmack, E., Mize, G.J., Morris, D.R., Garvik, B.M., and Yates, J.R. III (1999). Direct analysis of protein complexes using mass spectrometry. *Nature Biotechnol.* 17:676-682.

Lipp, M., Anklam, E., and Stave, J. (2000). Validation of an immunoassay for the detection and quantification of Roundup-Ready reference material. *J. AOAC Int.* 83:99-127.

Lipton, C.R., Dautlick, X., Grothaus, G.D., Hunst, P.L., Magin, K.M., Mihaliak, C.A., Ubio, F.M., and Stave, J.W. (2000). Guideline for the validation and use of immunoassays for determination of introduced proteins in biotechnology enhanced crops and derived food additives. *Food Agric. Immunol.* 12:153-164.

MacCallum, R.M., Martin, A.C., and Thornton, J.M. (1996). Antibody-antigen interactions: Contact analysis and binding site topography. *J. Mol. Biol.* 262:732-745.

McCafferty, J., Griffiths, A.D., Winter, G., and Chiswell, D.J. (1990). Phage antibodies: Filamentous phage displaying antibody variable domain. *Nature* 348:552-554.

McDonnell, R.E., Clark, R.D., Smith, W.A., and Hinchee, M.A. (1987). A simplified method for the detection of neomycin phosphotransferase activity in transformed tissue. *Plant Mol. Biol. Rep.* 5:380-386.

McNeal, J. (1988). Semi-quantitative enzyme-linked immunosorbent assay of soy protein in meat products: Summary of collaborative study. *J. AOAC Int.* 71:443.

Mihaliak, C.A. and Berberich, S.A. (1995). Guidelines to the validation and use of immunochemical methods for generating data in support of pesticide registration. In *Immunoanalysis of Agrochemicals* (pp. 56-69), J.O. Nelson, A.E. Karu, and B. Wong, eds. American Chemical Society, Washington, DC.

Miles, L.E.M. and Hales, C.N. (1968). Labelled antibodies and immunological assay systems. *Nature* 219:186-189.

Ornstein, L. (1964). Disc electrophoresis—I. Background and theory. *Ann. NY Acad. Sci.* 121:31-344.

Pluzek, K.-J. and Ramlau, J. (1988). Alkaline phosphatase labelled reagents. In *CRC Handbook of Immunoblotting of Proteins*, Volume 1 (pp. 177-188), *Technical Descriptions*, O.J. Bjerrum and N.H.H. Heegaard, eds. CRC Press, Boca Raton, FL.

Poulsen, F. and O.J. Bjerrum (1991). Enzyme immunofiltration assay for human creatine kinase isoenzyme MB. In *Immunochemistry of Solid-Phase Immunoassay* (pp. 75-82), J.E. Butler, ed. CRC Press, Boca Raton, FL.

Pound, J.D. (1998). *Immunochemical Protocols*, Second Edition. Humana Press, Totowa, NJ.

Rittenburg, J.H. and Dautlick, J.X. (1995). Quality standards for immunoassay kits. In *Immunochemistry of Solid-Phase Immunoassay* (pp. 44-52), J.E. Butler, ed. CRC Press, Boca Raton, FL.

Rogan, G.J., Dudin, Y.A., Lee, T.C., Magin, K.M., Astwood, J.D., Bhakta, N.S., Leach, J.N., Sanders, P.R., and Fuchs, R.L. (1999). Immunodiagnostic methods for detection of 5-enolpyruvylshikimate-3-phosphate synthase in Roundup Ready soybeans. *Food Control* 10:407-414.

Rogan, G.J., Ream, J.E., Berberich, S.A., and Fuchs, R.L. (1992). Enzyme-linked immunosorbent assay for quantitation of neomycin phosphotransferase II in genetically modified cotton tissue extracts. *J. Agric. Food Chem.* 40:1453-1458.

Sambrook, J. and Russel, D. (2000). *Molecular Cloning: A Laboratory Manual*, Third Edition, Cold Spring Harbor Laboratory, Cold Spring, NY.

Stave, J.W. (1999). Detection of new or modified proteins in normal foods derived from GMO—future needs. *Food Control* 10:367-374.

Sutula, C.L. (1996). Quality control and cost effectiveness of indexing procedures. *Adv. Botan. Res.* 23:279-292.

Thorpe, G.H.G. and Kricka, L.J. (1986). Enhanced chemiluminescent reactions catalyzed by horseradish peroxidase. *Meth. Enzymol.* 133:331-352.

van Duijn, G., van Biert, R., Bleeker-Marcelis, H., Peppelman, H., and Hessing, M. (1999). Detection methods for genetically modified crops. *Food Control* 10: 375-378.

Wedemayer, G.J., Patten, P.A., Wang, L.H., Schultz, P.G., and Stevens, R.C. (1997). Structural insights into the evolution of an antibody combining site. *Science* 276:1665-1669.

Williams, E. (2002). Advances in ultrahigh-field NMR technology: Proteomics and genomics. *Amer. Genomic/Proteomic Technol.* 2:28-31.

Yallow, R.S. and Berson, S.A. (1959). Assay of plasma insulin in human subjects of immunological methods. *Nature* 184:1648-1649.

Chapter 6

Protein-Based Methods: Case Studies

James W. Stave

INTRODUCTION

The use of modern agricultural biotechnology (agbiotech) methods has led to the development of novel crops that exhibit unique agronomic characteristics. These crops have novel pieces of DNA inserted into their genome that code for the production of unique proteins. The agbiotech DNA and protein can be found in a variety of plant tissues, including seeds and grain, and foods derived from them. A number of countries have enacted laws mandating that foods containing agbiotech ingredients above a specified threshold must be labeled as such. The threshold concentration specified in these regulations is defined in terms of the percentage of genetically modified organisms (GMO) present in the food or ingredient.

FOODS: PROCESSED AND FINISHED

The concentration of GMO in a sample is calculated by determining the concentration of agbiotech protein or DNA in the sample and estimating the percent GMO based on assumptions regarding protein expression levels and gene copy numbers in specific crops and tissues. Natural biological variability associated with protein expression and gene copy number contribute to the overall variability of the estimate of percent GMO (Stave, 2002; Van den Eede et al., 2002). For example, the difference in Cry1Ab protein concentration levels

Thanks to Dale Onisk for development of antibodies and George Teaney and Michael Brown for method development. It has been my pleasure to work with you.

between unique corn events expressing the protein is so great that quantification of an unknown sample using a Cry1Ab protein method is not possible. Similar problems can be found with DNA methods. For example, the estimate of percent GMO in an unknown corn sample differs by twofold depending on whether a biotech gene detected in the sample is assumed to derive from a plant that contains only a single biotech gene or a plant containing two biotech genes "stacked" together. Biological factors such as these significantly complicate the determination of percent GMO of unknown samples.

Biological variability determines the degree to which reality deviates from our assumptions regarding the correlation between target analyte concentration (protein or DNA) and percent GMO. In addition to biological variability, factors such as extraction efficiency, matrix effects, and target protein conformation impact the accuracy and precision of a method (Stave, 1999; Lipton et al., 2000). Figure 6.1 illustrates the issues associated with characterizing the extraction efficiency of Cry1Ab protein from corn flour. The figure shows the response, in a Cry1Ab enzyme-linked immunosorbent assay (ELISA), of sequential extracts of a single sample of flour made from corn expressing the Cry1Ab protein. The data demonstrate that the greatest amount of Cry1Ab is present in the first extract and that additional Cry1Ab protein is present in subsequent extracts.

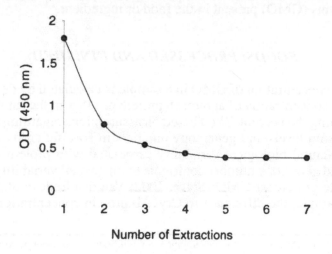

FIGURE 6.1. Extraction efficiency and ELISA reactivity of sequential extracts of Cry1Ab corn flour

Sample preparation procedures are a balance between efficiency and utility. It is not necessary to extract 100 percent of a protein to accurately determine its concentration in the sample. What is necessary, however, is that the amount of protein that is extracted by the procedure is consistent. Likewise, it is not necessary to know what percent of the protein is extracted by the procedure as long as the method is calibrated to a reference material of known concentration.

The effect of a matrix on method performance can be seen in Figure 6.2, where purified Cry1Ac protein is diluted in various concentrations of a negative leaf extract. The presence of low concentrations of leaf extract (20 mg/mL) has little effect on the method, whereas high concentrations significantly inhibit the response to Cry1Ac. Analogous to extraction efficiency, it is not necessary that methods be completely free from matrix effects as long as each sample contains a consistent amount of matrix that does not vary in composition, and the method is calibrated using a reference material of the same matrix and known concentration.

When testing processed food fractions using immunoassays, it is important to characterize the capacity of the antibody used in the test to bind to the form of the protein in the sample. Food processing procedures can alter the conformation of a protein in a way that the antibody can no longer bind, and it is important that the method used for

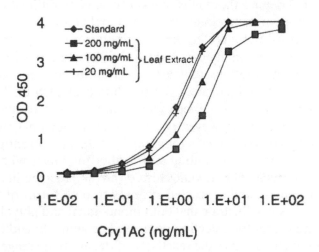

FIGURE 6.2. Effect of negative leaf matrix on ELISA reactivity to purified Cry1Ac protein standards.

analysis of a particular food substance be validated for applicability. Antibodies and test methods that can detect denatured CP4 5-enolpyruvylshikimate-3-phosphate synthase (EPSPS) in toasted soymeal have been previously reported (Stave, 2002). These same antibodies also detect CP4 EPSPS protein in other food substances prepared from soybean, including protein isolate, soy milk and tofu (Stave et al., 2004).

The importance of characterizing the reactivity of a method for the form of the protein found in a particular food matrix is further illustrated in the case of StarLink corn. Due to the inadvertent contamination of food with this corn, which had been approved for use in animal feed in the United States only, a considerable amount of work was expended to determine whether the Cry9c protein was present in foods prepared from StarLink. Some of the immunoassay methods available at the time employed monoclonal antibodies, whereas others used polyclonal antibodies. Using the antibodies from these methods, it was demonstrated that the polyclonal antibodies could detect low levels of Cry9c in some food fractions, whereas the monoclonals could not. These observations led to a general discussion of the utility of these two types of antibodies in immunoassays designed to be used with highly processed food fractions.

Figure 6.3 depicts the results of a study demonstrating the reactivity of a variety of monoclonal and polyclonal antibodies to a protein that was processed by boiling for 30 minutes or boiling for 30 minutes and then additionally exposed to a mild treatment with base. In the experiment, dilutions of the various antibodies, in decreasing concentration, were incubated within columns of each ELISA plate. Most of the monoclonal antibodies tested did not bind to either preparation of the protein, however, two different monoclonal antibodies (indicated by arrows) bound strongly only to the protein antigen once it had been subjected to boiling and further treatment with NaOH. One of the polyclonal antibodies tested (arrow), reacted in a similar fashion, whereas another reacted strongly to both forms of the antigen. These results illustrate that both monoclonal and polyclonal antibodies are useful for detecting processed proteins. In addition, it is important to characterize the reactivity of the antibody reagents used in a method to determine applicability for use with any given food type.

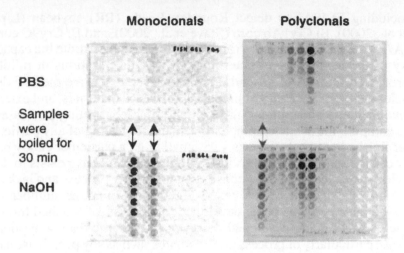

FIGURE 6.3. ELISA response of monoclonal and polyclonal antibodies to a boiled protein antigen further treated with mild NaOH

In order to calibrate a method designed to be used with a variety of different food types, it is necessary to prepare reference materials, containing known concentrations of agbiotech ingredients, according to the same procedures that are used to make the foods to be tested by the assay. These reference materials are extracted in the same way as the samples, and therefore matrix effects and extraction efficiency are normalized. Because the reference materials are prepared according to the same process as the food sample being tested, the form of the protein in the sample and the reference is the same. Once the method has been calibrated using reference materials of known concentration, it is possible to assign concentration values to standards that are run with every assay to construct a standard curve. These standards can be something other than the reference material, e.g., purified protein, as long as the response generated by the standard is the same as the response of the reference material having the concentration assigned to the standard.

Once a method has been developed, it is important to validate the performance through independent, outside parties. A number of ELISA methods for detection and quantification of agbiotech ingredients have successfully undergone extensive, worldwide validation, including ELISAs to detect Roundup Ready (RR) soybean (Lipp et al., 2000), Bt Cry1Ab corn (Stave et al., 2000), and Bt Cry9C corn (AACC, 2000). The results from these studies demonstrate the capacity of the technology to determine GMO concentrations in mildly processed samples of ground grain. However, extensive method development, the necessity of preparing reference materials, and extensive method validation are difficult and time-consuming processes. In addition, the actual commercial demand for testing of any particular finished food substance is very small (e.g., a particular candy bar or canned sausage), and thus little commercial incentive exists to develop such a test. The time-consuming and costly process and lack of commercial incentive have acted together to limit the number of quantitative methods being developed for analysis of finished foods. Methods are being developed however, for specific testing applications, particularly of process intermediates such as soy protein isolate or defatted flake, where a relatively large volume of regular testing of the same sample type occurs.

GRAIN AND SEED

Although little effort is being expended to develop quantitative protein methods for finished foods, immunoassay methods for detection of biotech traits in grain and seed are in widespread use. Rapid, lateral-flow strip tests are used for these applications. Their ease-of-use, rapid time-to-result, and low cost are ideally suited for detecting biotech commodities in containers or vessels using statistical techniques designed to determine whether the concentration of biotech in a consignment is above or below a specified threshold with high statistical confidence (Stave et al., 2004). The sampling and testing protocol employed can be tailored to meet the specific risk tolerances of the buyer and seller involved in a given transaction.

When a consignment containing relatively low concentrations of biotech grain is sampled, the probability that the sample actually contains a biotech kernel, bean, or seed is determined by factors such as the size of the sample and the concentration and distribution of biotech kernels in the load. Because some uncertainty associated with sampling in this fashion is always present, there is a risk that the test protocol will result in a misclassification of the load, i.e., cause the load to be rejected when it should be accepted or accepted when it should be rejected. These two types of risk are referred to as seller's and buyer's risk, respectively, and the relationship between them is illustrated in Figure 6.4.

Due to the heterogeneous distribution of biotech seeds (or beans or kernels, etc.) within a load, some probability exists that a sample from the load will contain a concentration that is higher or lower than the true concentration. Using a sample size of 1,000 seeds, the probability of accepting a load containing various percentages of biotech is defined by the operating characteristic (OC) curve shown in the figure. The buyer's and seller's risks for a load containing a true concentration of 1 percent biotech units, randomly distributed in the load, is indicated by the area of the figure between the OC curve and the line representing 1 percent. Using these principles, different sampling and testing protocols have been developed to meet the demands of specific markets.

An example of how this scheme is being used on a wide scale is testing for RR soybeans in Brazil. Brazil is the second leading producer of soybeans in the world, and existing Brazilian regulations prohibit the cultivation of agbiotech crops. In neighboring Argentina, approximately 95 percent of the soybean crops under cultivation consist of RR varieties. It has been estimated that as much as 30 percent of the soybean crops under cultivation in the southern Brazilian states, nearest Argentina, are RR despite the current ban, due to smuggling of

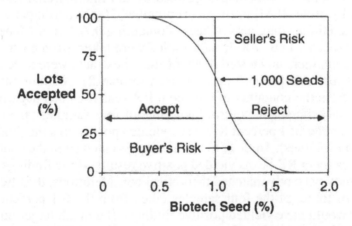

FIGURE 6.4. Operating characteristic curve illustrating buyer's and seller's risk (*Source:* Reprinted from *The Journal of AOAC International*, 2001, Vol. 84, pp. 1941-1946. Copyright 2001 by AOAC International.)

beans across the border. At present a significant market for "non-GMO" beans exists, and Brazilian producers are able to capitalize on this opportunity, provided the beans that they sell are indeed non-GMO.

To ensure the quality of the product that they ship, Brazilian producers have widely implemented testing protocols using rapid, lateral-flow strip tests and threshold sampling. The most common protocol consists of collecting a single representative sample of 700 beans from a load, grinding the beans to a powder, extracting with water, and reacting the extract with a strip test. If the test is positive, the load is rejected. It is important to note that the USDA's Grain Inspection, Packers and Stockyards Administration has verified that the strip test is capable of detecting 1 RR bean in 1,000, but the sample size is limited to 700 to ensure that no false negatives result. Assuming that the manner in which the representative sample is taken overcomes non-uniform distribution of the biotech beans in the load and accurately reflects a random sample, then, using the OC curve for this protocol, it can be calculated that the probability of detecting (and rejecting) a load consisting of 1 percent RR beans is 99.9 percent. The probability of detecting a load at 0.1 percent is 50 percent, i.e., the buyer's and seller's risks are equal.

The results of such an analysis are depicted in Figure 6.5. At a detection methods workshop organized by the International Life Sciences Institute (ILSI) in Brazil (September, 2002), participants were provided three containers of beans containing 0, 0.1, and 1.0 percent RR soybeans. Four 700-bean samples were taken from each of the three containers and tested. Each of the 12 extracts were tested with two strips (except the 0 percent from sample 2). All four samples taken from the container of 0 percent RR beans yielded negative results (only the single control line), and all four samples taken from the container of 1 percent RR beans yielded positive results (both test and control lines). Three of the four samples taken from the container of 0.1 percent RR beans yielded positive results. If the findings from this study perfectly matched theoretical considerations, then two, instead of three, of the four samples taken from the 0.1 percent container would have yielded positive findings. If a much larger number of samples were taken from the same container of 0.1 percent RR beans, then the number of negative and positive analyses would be more equal.

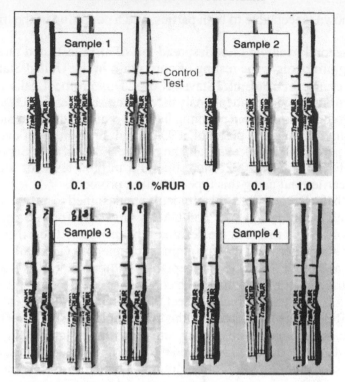

FIGURE 6.5. Strip-test results using RR soybeans from ILSI Workshop on Detection Methods for GMOs in the Food Chain, September 2002, Rio de Janeiro, Brazil. Results illustrate sample heterogeneity when buyer's and seller's risks are equal at 0.1 percent GMO.

It is common for inexperienced analysts to misinterpret the single negative result at 0.1 percent RR (sample 2), as a false negative result when, in reality, it is the expected outcome given the concentration of RR beans and sampling heterogeneity. The test strip was not falsely negative because that particular sample did not contain a positive bean. Similarly, it would be incorrect to view the third positive result at 0.1 percent as a false positive the strip accurately detected the presence of an RR soybean. The presence and detection of a third sample containing an RR bean was simply a consequence of sampling heterogeneity. These findings highlight the importance of designing sampling and testing protocols in a way that the risks of misclassifying the sample are known, acceptable to both parties, and a condition of the transaction.

A second example of widespread use of strip tests and threshold testing protocols is the testing of corn grain in the United States for the presence of StarLink (Stave, 2001). This protocol calls for the collection and independent analysis of three samples of 800 kernels from a consignment, and rejecting the load if any of the three samples is positive. Using this protocol, a 95 percent probability of detecting 0.125 percent StarLink exists. In the two-year period from September 2000 to September 2002, more than 3.5 million tests for StarLink were carried out using this type of testing protocol.

Both the RR and StarLink protocols are designed to ensure that the products being tested are "non-GMO," and therefore no consideration of seller's risk is in the design of the protocol. The existence of even a single kernel of biotech grain is sufficient to violate non-GMO status and render the load unacceptable. Although the 700-bean protocol used in Brazil provides high confidence of rejecting a load containing 1 percent, it will also detect 0.1 percent RR 50 percent of the time. Rejection of 50 percent of the loads containing 0.1 percent may not be acceptable in all testing situations. In situations in which biotech grains have been approved by government authorities, appreciable concentrations of biotech grain in a load may be tolerated. For example, RR soybeans have been approved for use in Europe. Regulations mandate that foods containing 1 percent or greater must be labeled as containing GMO. In this case, it is possible to imagine a transaction in which the buyer wants high confidence that the beans being purchased contain less than 1 percent RR to avoid labeling. At the same time, the seller wants reasonable confidence that if the load indeed contains less than 1 percent RR, a high probability exists that the load will be accepted. Figure 6.6 illustrates an example of a sampling and testing protocol that could be useful under such a scenario to provide acceptable assurances to both parties.

A limitation of this technology is the difficulty of testing for all available biotech events using a single sample. At present, only one significant biotech variety of soybean is sold in the commercial marketplace, and therefore simply determining the concentration of RR beans in a sample is equivalent to determining percent GMO. In corn however, many different biotech events are present, each expressing their respective biotech proteins at significantly different concentra-

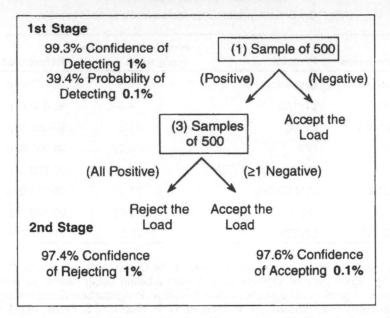

FIGURE 6.6. Threshold testing protocol providing high confidence that 1 percent will be rejected and 0.1 percent will be accepted. Calculations were performed using SeedCalc5. (*Source:* Remund et al., 2001. Reprinted from *Seed Science Research*, June 2001, Vol. 11, No. 2, pp. 101-120. Copyright 2001 by CABI Publishing.)

tions (Table 6.1). It is technically possible to test for multiple biotech events in a single sample. Indeed, this is a common practice in the cotton industry, where seed producers regularly test seeds to ensure they contain expected traits (Figure 6.7). However, to use the threshold testing protocol, it is necessary to limit the maximum number of kernels in a sample so that the presence of only a single biotech kernel will always give a positive response. When performing threshold testing for multiple biotech events in a single sample, it is necessary to limit the number of kernels so that the lowest expressing event will always be detected. However, such a constraint significantly limits the sensitivity of the methods for high-expressing events. This constraint on sensitivity, and the large and ever-changing list of biotech events, have curtailed efforts to develop methods to detect multiple biotech events in a single sample of corn grain.

TABLE 6.1 Biotech protein expression levels in corn kernels

Corn Event	Protein	Protein Expression (µg/g fresh weight)	Reference‡
MON 810	Cry1Ab	0.3	96-017-01p
BT11	Cry1Ab	4.8	95-195-01p
CBH-351	Cry9C	15.0	97-265-01p
T25	PAT	0.12	94-357-01p
GA21	Modified corn EPSPS	3.2	97-099-01p
NK603	CP4 EPSPS	11.0	00-011-01p
1507	Cry1F	3.0†	00-136-01p
MON 863	Cry3Bb	70.0	01-137-01p

Source: Stave, J. W., Brown, M. C., Chen, J., McQuillin, A. B., and Onisk, D. V. (2004). Enabling gain distribution and food labeling using protein immuno-assays for products of agricultural biotechnology. In *Agricultural Biotechnology: Challenges and Prospects*, M. M. Bhalgat, W. P. Ridley, A. S. Felsot, and J. N. Seiber, eds. (pp. 69-82). American Chemical Society, Washington, DC.
†Estimated by authors.
‡Petitions of Nonregulated Status Granted by USDA APHIS.

DISCUSSION

Biological variation of protein expression within and between events, and low-level expression impose limits on the capacity of protein methods to quantify GMO in finished foods. The development of such methods is time-consuming and costly and requires the use of reference materials for each food matrix tested. Food companies interested in selling products that comply with GMO labeling laws have found that a part of the solution is to test incoming ingredients and then control the ingredient in their processes in a way that the concentration is known.

Threshold testing using rapid, lateral flow strip tests has been implemented successfully on a large scale for corn grain and soybeans. Further deployment of the technology will require increasing the capacity to detect multiple traits in a single sample. Early biotech events,

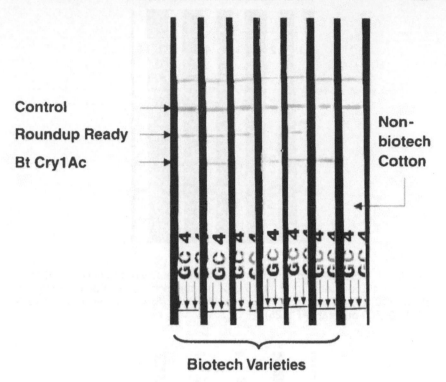

FIGURE 6.7. Multievent strip test for Roundup Ready and Bt Cry1Ac cotton seed

such as Bt 176 and GA21 corn, which imposed unique problems to protein tests, are being replaced in the marketplace by such new events as NK603, 1507 and MON 863. These events have relatively high levels of biotech protein expression, and rapid methods have been successfully developed for these events (Figures 6.8 and 6.9). Agencies such as the U.S. Environmental Protection Agency are making the existence of test methods a condition of biotech product registration, insuring that detectability is a consideration of biotech crop development. The disappearance of problematic events and the introduction of new high-expressing events that are designed to be detectable improve the prospects of developing rapid threshold testing protocols for multiple events in a single sample.

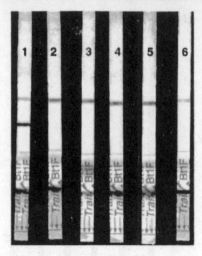

FIGURE 6.8. Specificity of strip test for detection of Cry1F protein in event 1507 corn: (1) 100 percent Cry1F (1507); (2) nonbiotech; (3) 100 percent Cry1Ab (MON 810); (4) 100 percent CP4 EPSPS (NK603); (5) 100 percent Cry9C (CBH-351); (6) 100 percent PAT (T25)

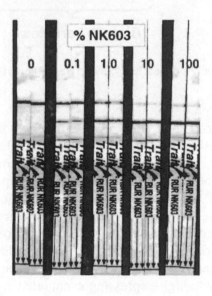

FIGURE 6.9. Sensitivity of strip-test method for RR corn event NK603 expressing CP4 EPSPS

REFERENCES

American Association of Cereal Chemists (AACC) (2000). Method 11-20. StarLink corn in corn flour and corn meal—ELISA method. In *Approved Methods of the AACC,* Tenth Ed. AACC, St. Paul.

Lipp, M., Anklam, E., and Stave, J.W. (2000). Validation of an immunoassay for detection and quantification of a genetically modified soybean in food and food fractions using reference materials: Interlaboratory study. *J. AOAC Int.* 83(4): 919-927.

Lipton, C.R., Dautlick, J.X., Grothaus, G.D., Hunst, P.L., Magin, K.M., Mihaliak, C.A., Rubio, F.M., and Stave, J.W. (2000). Guidelines for the validation and use of immunoassays for determination of introduced proteins in biotechnology enhanced crops and derived food ingredients. *Food Agricultural Immunol.* 12:153-164.

Remund, K., Dixon, D.A., Wright, D.L. and Holden, L.R. (2001). Statistical considerations in seed purity testing for transgenic traits. *Seed Sci. Res.* 11: 101-120.

Stave, J.W. (1999). Detection of new or modified proteins in novel foods derived from GMO—Future needs. *Food Control* 10:367-374.

Stave, J.W. (2001). Testing for StarLink: Analytical method availability and reliability. In *StarLink: Lessons Learned* (pp. 51-58), FCN Publishing, eds. CRC Press, Washington, DC.

Stave, J.W. (2002). Protein immunoassay methods for detection of biotech crops: Applications, limitations, and practical considerations. *J. AOAC Int.* 85(3):780-786.

Stave, J.W., Brown, M.C., Chen, J., McQuillin, A.B., and Onisk, D.V. (2004). Enabling grain distribution and food labeling using protein immunoassays for products of agricultural biotechnology. In *Agricultural Biotechnology: Challenges and Prospects.* M.M. Bhalgat, W.P. Ridley, A.S. Felsot, and J.N. Seiber, eds. (pp. 69-82). American Chemical Society, Washington, DC.

Stave, J.W., Magin, K., Schimmel, H., Lawruk, T.S., Wehling, P., and Bridges, A.R. (2000). AACC collaborative study of a protein method for detection of genetically modified corn. *Cereal Foods World* 45(11):497-500.

Van den Eede, G., Kay, S., Anklam, E., and Schimmel, H. (2002). Analytical challenges: Bridging the gap from regulations to enforcement. *J. AOAC Int.* 85(3): 757-761.

Chapter 7

DNA-Based Methods for Detection and Quantification of GMOs: Principles and Standards

John Fagan

INTRODUCTION

This chapter discusses the genetics-based methods currently used worldwide to detect and quantify genetically modified organisms (GMOs) in foods and agricultural products. The perspective presented emerges from the experience our company, Genetic ID, has gained over a period of more than six years, during which we have developed analytical methods for GMOs and applied these in service to the food and agricultural industries, and to governments around the world.

At present more than 36 countries have established laws or regulations requiring the labeling of foods and agricultural products that are genetically modified or that contain genetically modified ingredients. These include the European Union (15 countries), Switzerland, Norway, Indonesia, Poland, China (including the Hong Kong special administrative region), Japan, Latvia, Croatia, Russia, Australia, Bolivia, New Zealand, Saudi Arabia, South Korea, Thailand, Czech Republic, Israel, Mexico, Philippines, and Taiwan. In essence, all global centers of economic power, except North America, have established such laws. These laws, as well as international accords, such as the Cartagena Protocol on Biosafety make it necessary for governments and industry to monitor and disclose the presence of GMOs in food and agricultural products.

In order to implement these regulations, government laboratories must possess the capacity to test for GMOs. The food and agricultural industries also need to have access to reliable testing capabilities. This testing capacity enables them not only to comply with labeling regulations but also to respond to consumers' demands for transparency regarding the presence of GMOs in food.

Analytical methods are needed that can detect and quantify GMOs at all levels of the food chain. Gene modification (also called recombinant DNA methods or gene splicing techniques) introduces new genetic information, new DNA sequences, into the genome of an organism. As illustrated in Figure 7.1, once introduced into the genome, the transgenic (also called genetically modified) DNA reprograms the cells of the recipient organism to produce new mRNA and proteins. The transgenic proteins confer new characteristics or functions upon the organism. In principle, GMO detection methods could target transgenic DNA, mRNA, proteins, or novel biological molecules uniquely

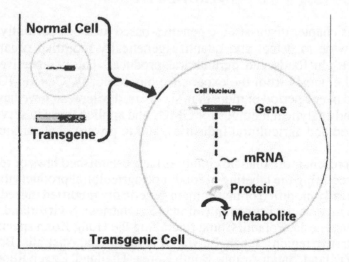

FIGURE 7.1. Targets for GMO analysis. The creation of a transgenic organism involves the insertion of new genetic information into the genome of the host organism. This new DNA reprograms the cells of the organism to produce new mRNA molecules, which in turn guide the synthesis of new proteins. These new proteins confer new properties onto the cell. For example, they can catalyze the synthesis of metabolites that are not normally present in the host organism. In principle, the transgenic organism can be distinguished from the host organism by the presence of any one of these new molecules, the transgenic DNA, the mRNAs synthesized from the transgenic DNA template, the transgenic protein, and the new biomolecules synthesized by the transgenic protein. In practice, practical detection methods have focused on the transgenic proteins and genes.

produced in the transgenic organism as a direct or indirect result of the function of transgenic proteins. However, to date, GMO analytical methods have focused almost exclusively on DNA and protein.

GMO tests targeting proteins employ immunological methods including both the enzyme-linked immunosorbent assay (ELISA) and the lateral flow test formats (Anklam, Gadani, et al., 2002; Lipp et al., 2000; Lipton et al., 2000; Stave, 1999, 2002). Because they require minimal processing of the sample, immunotests can be completed quite quickly (Lipton et al., 2000). Moreover, in the lateral flow format, immunotests are convenient and easy to carry out and do not require sophisticated equipment. This format is particularly useful in the field, where it can be used to rapidly and inexpensively screen truckloads of soy or maize at the grain handling facility for single, genetically modified traits.

This chapter will focus on tests for GMOs that detect and quantify those DNA sequences that have been introduced into the organism during the process of gene modification. To date, GMO tests targeting genes have relied almost exclusively on the polymerase chain reaction (PCR) analytical platform (Anklam, Gadani, et al., 2002; Greiner et al., 1997; Hemmer, 1997; Hemmer and Pauli, 1998; Hoef et al., 1998; Hübner, Studer, Hufliner, et al., 1999; Hübner, Studer, and Lüthy, 1999a; Jankiewicz et al., 1999; Lipp et al., 1999; Meyer, 1999; Schweizerischers Lebensmittelbuch, 1998; Shirai et al., 1998; Vaitilingom et al., 1999, Vollenhofer et al., 1999; Wassenegger, 1998; Wiseman, 2002). PCR uses biochemical processes to scan through a sample of DNA and locate one or more specific DNA sequences called *target sequences*. If the target sequence is present, the PCR process amplifies it billions of times, making it possible to detect target sequence with high sensitivity and quantify the proportion of DNA molecules in the sample that contain that target.

THE POLYMERASE CHAIN REACTION

The key elements in the PCR process are as follows:

1. "Primers"—small DNA molecules whose sequences correspond to the target sequence

2. A heat-stable DNA polymerase—typically Taq polymerase, which synthesizes new copies of the target sequence in a manner that is dependent upon the interaction of primers with these target sequences

3. A thermocycler—an apparatus that can be programmed to carry the contents of the PCR reaction vessel through multiple, precisely controlled temperature cycles.

The PCR amplification process is illustrated in Figure 7.2. Using this system, a specific target sequence can be located within a DNA sample and amplified billions of times. The process is initiated by combining all reaction components in a sealed reaction vessel. Once sealed, different steps in the PCR cycle are initiated by shifting the reaction vessel from one temperature to another without adding more components to the vessel. Each cycle of temperature conditions doubles the number of target DNA sequences in the reaction vessel. The

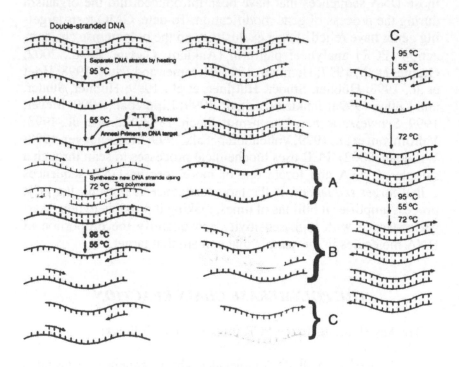

FIGURE 7.2. The polymerase chain reaction

synthesis of these DNA molecules is monitored either in real time, cycle by cycle, or at a single point in time after several cycles have occurred.

In the example presented in Figure 7.2, the PCR cycle consists of three temperatures, 95°C, 55°C, and 72°C. First, the reaction vessel is shifted to 95°C to denature the double-stranded DNA template molecules (class A molecules in Figure 7.2), separating them into single strands. The vessel is then shifted to 55°C, where the primer molecules can interact or "anneal" with the DNA template if sequences complementary to the primers are present in the DNA template. The result is the formation of stable hybrids between primers and targets. The vessel is then shifted to 72°C, the optimal temperature for Taq polymerase to catalyze DNA synthesis. At this temperature, Taq polymerase uses the primer hybridized to the DNA template as the starting point for synthesizing a DNA strand complementary to the DNA template molecule. The result is two double-stranded molecules, as illustrated in Figure 7.2, with the new DNA strands having one end delimited by one of the primers (class B molecules in Figure 7.2).

A second cycle of denaturation, primer annealing, and DNA synthesis produces more class B DNA molecules, but also a third class (class C) in which both ends of the molecule are delimited by primer sequences. In subsequent cycles, the class B molecules increase in number arithmetically, whereas the class C molecules increase exponentially, rapidly becoming the predominant product. Starting with one (1) class C molecule, the first cycle results in 2, the second 4, the third 8, the fourth 16, and so on. In ten cycles, 1,024 molecules are generated, in 20, more than 1 million, and in 30, more than 1 billion are generated.

The exponential nature of this amplification reaction is illustrated by curve A in Figure 7.3. In theory, production of PCR products should proceed exponentially. In actual practice, however, PCR does not proceed exponentially for an unlimited period of time. Instead, it reaches a plateau, as shown in curve B. This occurs because certain reaction components, most critically deoxyribonucleotide triphosphates (dXPTs), primers, and Taq polymerase, become limiting as the reaction proceeds.

Figure 7.4 plots the time course of the polymerase chain reaction as a function of target DNA concentration using log-linear coordinates. We see that the shape of the reaction profile is essentially inde-

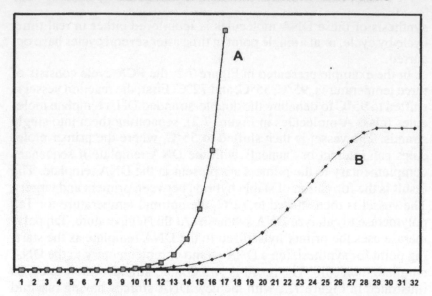

FIGURE 7.3. Comparison of exponential amplification and the typical amplification profile of PCR. Curve A presents a theoretical exponential amplification profile. The initial phase of each PCR reaction is exponential. However, as Taq polymerase, dXTPs, and primers become limiting, the reaction plateaus,resulting in a profile similar to that of curve B.

pendent of the target DNA concentration. What changes is the number of cycles of PCR required to generate that profile. For example, a reaction containing DNA that is 10 percent genetically modified will plateau after fewer cycles than a reaction containing DNA that is 1 percent genetically modified. This phenomenon is exploited in quantitative PCR methods, as will be discussed in later sections.

PCR Detection Systems

In simplest terms, the PCR process asks a very specific question: Does the sample DNA contain sequences that are complementary to the primer set added to the PCR reaction? If present, billions of copies of that sequence, called *amplicons,* are generated through the PCR process. If it is absent, no amplification takes place. Detection of whether the amplification process has occurred is accomplished in sev-

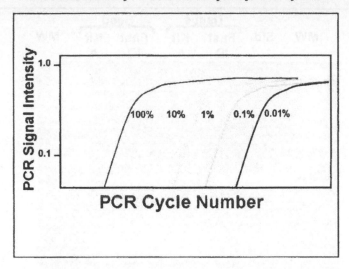

FIGURE 7.4. Log-linear reaction profiles for PCR reactions. The profiles of PCR reactions containing various concentrations of genetically modified target sequences are all similar in shape but differ in the number of PCR cycles required to achieve a given intensity of PCR signal.

eral ways. Most commonly, amplicons are detected directly through electrophoretic fractionation of the PCR reaction products, followed by staining with an intercalating dye, such as ethidium bromide. A typical electrophoretic result is presented in Figure 7.5. Hybridization methods using either radioactive or fluorescent probes have also been used to detect amplicons directly. In recent years, the microarray format for hybridization detection (Kok et al., 2002; Shena, 2000) has supplanted earlier use of dot-blot and Southern blot formats. Immunolinked detection methods have also been employed for direct detection of amplicons (Demay et al., 1996; Taoufik et al., 1998). In these methods, amplicons are labeled with digoxygenin or other immunoreactive moeities, these are then detected using alkaline phosphatase-linked antibodies specific for the immunoreactive moiety (e.g., digoxygenin) and either colorimetric (Demay et al., 1996) or luminometric (Taoufik et al., 1998) substrates for alkaline phosphatase. Biosensors based on surface plasmon resonance (Feriotto et al., 2002; Mariotti et al., 2002; Minunni et al., 2001), piezoelectric, and electrochemical (Minunni et al., 2001) principles have

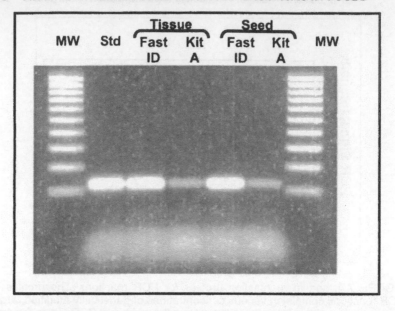

FIGURE 7.5. Typical electrophoretogram of PCR products. DNA was extracted from tomato tissue and seeds by two methods, and amplification was carried out with primers targeting the tomato *GAPDH* gene. MW indicates molecular weight markers; Std indicates PCR reaction products from reference DNA preparation.

also been used to detect PCR reaction products, although these methods have not been used routinely.

Indirect detection methods for the PCR amplification process, based on fluorescent dye systems, such as the TaqMan (Higuchi et al., 1992), fluorescence resonance energy transfer (FRET) (Wittwer et al., 1997), molecular beacon (Tyagi and Kramer, 1996), Scorpion (Whitecomb, 1999), and SYBR Green (Wittwer et al., 1997) systems, have been found to be very useful, particularly in quantitative PCR applications. The chemical and physical principles upon which these dye systems are based are all quite distinct, and can be understood in detail from the articles cited earlier. However, the underlying principle is the same for the first four systems: a fluorescent molecule is created for each PCR product (amplicon) generated during the PCR process. The SYBR Green system operates by a simpler, nonspecific mechanism. SYBR Green is a fluorescent dye whose fluorescence yield greatly increases in efficiency following intercalation with double-stranded amplicons. SYBR Green fluorescence increases as amplicon concentration increases. SYBR Green will interact with virtually any double-stranded DNA molecule, whether it is a bona fide amplicon, a primer dimer, or other artifact. Thus, use of this method carries with it increased risk.

The fluorescence-based methods mentioned previously can be used for endpoint PCR, however, their greatest advantage is found in their use in kinetic PCR methods, because the full reaction time course can be followed using a single-sealed PCR reaction vessel that is repeatedly integrated with a fluorescence detection system. Several companies, including Applied Biosystems, Bio-Rad, Roche, and Stratagene, produce thermocyclers that automate the collection of fluorescence data from PCR reaction vessels, making it possible to conveniently follow time courses of as many as 96 reactions in real time simultaneously as will be discussed in more detail in the following.

Applying PCR to GMO Analysis

PCR analysis of GMOs involves four steps: (1) sampling and sample preparation, (2) DNA purification, (3) PCR amplification and detection of reaction products, and (4) interpretation of results. These steps are discussed in detail in the following sections. It is worth pointing out that each step is critical to the accuracy, reliability, and practical utility of PCR-based GMO analysis. Discussions of GMO analysis often focus on the cutting edge of the procedure, namely the PCR process itself. Although the polymerase chain reaction is a highly important part of the GMO analytical process, much more mundane components of the process, such as sampling, sample preparation, and DNA purification, are equally critical to the overall utility of the procedure. For example, in most cases sample size turns out to be the factor that limits the sensitivity of the method, not the PCR process per se.

Sample Preparation and Sampling Statistics

For GMO analysis to be relevant to the real world, it is essential that the sample analyzed by PCR be representative of the lot of food or agricultural product of interest. The field sample must be representative of the product lot, and the analytical sample must be representative of the field sample. To achieve this, the field sample must be obtained in a manner that ensures representation from all parts of the lot. Statistical methods are used to define a sampling plan that yields a representative sample. Here we will discuss three specific sampling issues critically related to PCR testing.

Field Sample Size

The first critical sampling issue is field sample size. The field sample must contain a sufficient number of units to ensure that the analysis will be statistically robust at the limits of detection and quantification relevant to the assay. If the sample size is too small, the full power of PCR cannot be exploited. Typically, the limit of detection for PCR is in the range of 0.01 percent, which means that, in a sample containing 10,000 soybeans, PCR is capable of detecting the presence of a single GM bean. Clearly, in a sample of only 500 beans (a common sample size in some labs), the sensitivity inherent in PCR is not fully utilized.

The statistical significance achievable with such small sample sizes is extremely weak. For example, if a shipment of soybeans contains on average 0.1 percent GMO (one GM bean per 1,000), the probability that one would detect the presence of any genetically modified material in a sample of 500 beans is only 39 percent. That is, using the binomial distribution, one can calculate that the probability of drawing one genetically modified bean when a total of 500 beans are sampled, and when the true proportion is one genetically modified bean in 1,000, is only 39 percent. The probability that at least one GM bean would be drawn, if a total of 1,000 beans (200 grams) were sampled, is higher (63 percent), but this is still not adequate because analyses using samples of this size would miss the presence of GMOs 37 percent of the time. A 37 percent rate of false negatives is not acceptable. In contrast, analyses carried out using samples of 10,000 beans would detect the presence of GM material 99.99 percent of the time. We recommend that samples of this size (10,000 beans or kernels) be used when analyzing unprocessed commodities, such as soy, corn, and oilseed rape (canola).

For many applications, mere detection of the presence of GM material is insufficient—quantitative information is needed. Another kind of statistic is used in evaluating sample sizes appropriate for quantitative analysis. For this purpose, it is necessary to calculate, not the probability that a single GM bean/kernel will be present in the sample, but the probability that the actual GMO content of a given sample will be between two bounds, given the mean GMO content of the lot from which the sample is to be taken. Using a method based on the normal approximation of the binomial distribution, if the actual proportion of GM soybeans contained in a particular shipment was 0.1 percent, an 89 percent probability exists that the number of GM beans in a 2.5 kg (10,000-bean) sample would be within 50 percent of the actual value. In other words, an 89 percent probability exists that the number of GM beans in the 10,000-bean sample would be greater than 5 and less than 15 (0.05 percent to 0.15 percent) for a lot whose mean GMO content is 10 GM beans per 10,000 (0.1 percent). This is good statistical strength, compared to that obtained with samples of a few hundred beans. A 25 percent probability exists that the number of GM beans in a 400-bean sample would be within 50 percent of the mean GM content of the lot from which the sample was taken, and a 32 percent probability exists that the number of GM beans in a 700-bean sample will be within 50 percent of the mean GMO content.

To achieve a probability of 99 percent, with a confidence interval of 50 percent, a 7.5 kg (30,000-bean) sample would be required, and to achieve 99.9 percent probability, a 10 kg sample would be required. Similarly, larger samples would be required to narrow the confidence interval. With a sample of 15 kg (60,000 beans), the probability would be 95 percent that the number of GM beans in the sample would be within 25 percent of the true proportion (1 in 1,000). It would be necessary to increase the sample size to 40 kg in order to achieve a probability of 99.9 percent that the number of GM beans in the sample would be within 25 percent of the mean proportion.

From a statistical point of view, it would be ideal to use samples of these larger sizes, and such large samples are used for some highly specialized purposes in the agricultural industry, for instance, in testing for the carcinogen aflatoxin (Whitaker et al., 1995) and in testing the GMO status of seed to very high stringency. However, for routine GMO testing it would be prohibitively expensive to ship such samples, and prohibitively time-consuming to process them. Recognizing these practical considerations, we recommend for routine purposes sample sizes in the 2.5 kg to 3.0 kg range because they strike a judicious compromise between statistical, logistical, and economic considerations.

Analytical Sample Size

The total field sample should be ground and homogenized. Duplicate subsamples should then taken in a manner that ensures that the analytical sample will be representative of the field sample. DNA is isolated independently from the duplicate analytical subsamples for later PCR analysis. The key question is how large should the subsamples be to ensure that each is fully representative of the field sample? We have approached this question using statistical analysis based on estimates of the total number of particles per gram present in the ground sample. For example, we determined that 73,000 particles are present in 1g of maize kernels ground according to standardized procedures. Calculations based on the normal approximation to the binomial distribution indicate that, if the true or mean GMO content of the ground field sample is 0.01 percent, then the probability $(P) = 0.9439$ that the GMO content of a sample of 2 g will contain 0.01 percent ± 0.005 percent. That is, $P = 0.9439$ that the GMO content of the 2 g sample will be between 0.005 percent and 0.015 percent. The probability that a 1 g sample would be within this range would be $P = 0.8233$. In contrast, the probability that samples of 0.1 g and 0.2 g would be in this range would be $P = 0.3308$ and 0.4543, respectively. We conclude that samples of at least 1 g, and preferably 2 g, are necessary. Clearly, samples of 0.1 g to 0.2 g, which are used in many laboratories, are inadequate. Empirical studies confirm these conclusions. We have established 2 g as the standard for samples of ground grains and oilseeds. Using samples of these larger sizes is consistent with recommendations of the Comité Européen de Normalisation/International Organization for Standardization (CEN/ISO) working group on GMO testing as well (Whitaker et al., 1995).

It is worthwhile to note that the particle number per gram, mentioned previously for maize, is for routine analytical samples. The procedure used is designed for speed and efficiency, reducing 3 kg of maize to powder in less than 7 minutes. Statistical analyses as well as empirical studies indicate that this homogenization procedure is adequate to yield reproducibly uniform results when coupled with appropriate sample size. Much higher particle numbers are achieved for reference materials.

DNA Aliquot Size

The third sampling issue critical to PCR analysis relates to the amount of sample DNA introduced into the PCR reaction. A typical PCR reaction for GMO analysis will contain 100 ng of DNA. This corresponds to approximately 38,000 copies of the maize genome (Kay and Van den Eede., 2001; Van den Eede et al., 2002). This number makes it obvious that genome copy number places strict boundaries on the limits of detection and on the precision of PCR analysis. In this case, 0.1 percent GMO would correspond to 38 GM genomes out of 38,000, and 0.01 percent GMO would correspond to 3.8 GM genomes. Clearly, the absolute limit of detection for PCR reactions containing 100 ng maize DNA would be around 0.01 percent. At 0.1 percent, reasonably good reproducibility exists between duplicate PCR reactions. However, at 0.01 percent there would be considerable statistical variation in the number of GM genomes picked up in any given 100 ng sample, which would lead to significant variation among replicate PCR reactions.

DNA Purification

To gain reliable and informative results, purification procedures must produce DNA that is free from PCR inhibitors. In addition, DNA degradation must be minimized, and DNA yield must be sufficient to allow adequate analysis. DNA extraction kits purchased from scientific supply houses are unlikely to perform adequately for all sample types because food products vary greatly in their physical and chemical composition. Thus, it is essential to customize DNA extraction methods to achieve optimal results with each food matrix or agricultural product. Laboratories have approached this challenge in different ways. Some have developed their own stand-alone DNA purification systems with multiple modules that are optimized for different food matrixes. It is also common for laboratories to modify the protocols for commercial DNA extraction kits to better handle various foods. The following two basic procedures are commonly used for isolation of DNA from food and agricultural products for GMO analysis:

1. Sample disruption and DNA solubilization in hexadecyltri-methylammoniumbromide (CTAB), followed by recovery of DNA by precipitation (Murray and Thompson, 1980; Scott and Bendich, 1998).
2. Sample disruption and solubilization in solutions containing detergents and chaotropic agents, followed by binding of DNA to silica and recovery of DNA by elution in a low-salt buffer (Melzak et al., 1996; Vogelstein and Gillespie, 1979). In some cases these kits use spin columns that contain immobilized silica, in others magnetic silica particles are used to capture the DNA. Recently, other binding matrixes, such as magnetite (Davies et al., 1998) and charged polymers have come into use.

Figure 7.6 compares the performance of three DNA extraction methods. In this study, DNA was extracted from soybeans, from soy protein isolate, a soy derivative used as an ingredient in many food products, and from two multiingredient products that contain soy protein isolate. The quality of the DNA prepared using these different methods was then assessed by real-time quantitative PCR. PCR-active DNA was recovered successfully from soybeans and soy protein isolate using all three methods. However, the quality of the DNA preparations differed. Only one of the methods yielded DNA from the multi-ingredient products that was highly active in real-time PCR, illustrating the critical role that DNA extraction methods play in enabling the laboratory to deliver strong, accurate analytical results from a wide range of food products.

It should be pointed out that in this study the commercial DNA extraction kit was used exactly according to the procedure recommended by the manufacturer. However, most laboratories that use such kits routinely for GMO analysis have modified the commercial protocols to improve performance with various food matrixes. In some cases, the CTAB method has been used in series with such kits to separate DNA from PCR inhibitors more effectively and thereby recover PCR-active DNA from problematic food products (Zimmermann et al., 1998). A number of studies have been published describing DNA extraction methods developed for GMO testing and

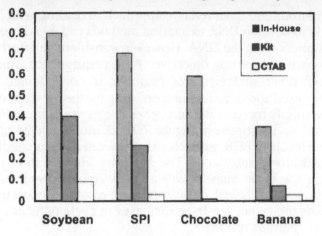

FIGURE 7.6. Comparison of three DNA extraction methods by real-time quantitative PCR analysis. DNA was prepared from four food samples using (1) an in-house method, (2) a commercial DNA extraction kit commonly used by various laboratories for analysis of genetically modified foods and agricultural products, and (3) the public domain CTAB method. The quality of the DNA was assessed by real-time PCR using primers specific for the glycine max (soy) *accC-1* gene. A standard amount of DNA (50 ng, quantified by absorbance at 260 nm) was introduced into each PCR reaction. PCR signals are reported relative to signals obtained with a standard of highly purified soy DNA. Reduced signals for chocolate and banana drinks for the in-house method are not due to the presence of inhibitors, but to the presence of DNA from other species derived from ingredients in these samples. SPI= soy protein isolate. Chocolate = chocolate-flavored, soy-based, multi-ingredient drink. Banana = banana-flavored, soy-based, multi-ingredient drink.

optimized protocols for the use of commercial kits for this purpose (Csaikl et al., 1998; Terry et al., 2002), and the work has been reviewed in detail (Anklam, Heinze, et al., 2002; Terry et al., 2002).

Dealing with PCR Inhibitors

The key role that DNA purification plays in establishing reliable GMO testing methods cannot be emphasized. Foods and agricultural products contain a host of numerous compounds that are inhibitory to PCR, and thus can interfere with the sensitivity and precision of GMO analysis. These include polysaccharides, polyphenols, proteins, Ca^{++}, Fe^{++}, and a host of other secondary metabolites and trace compounds.

The rigorous and most reliable approach to dealing with PCR inhibitors is to develop DNA extraction methods capable of separating such inhibitors from the DNA. However, considerable effort may be necessary to achieve this objective. As a consequence, some have sought out other strategies. For example, in conventional PCR, a commonly used approach for compensating for the presence of inhibitors is to simply increase the number of cycles used during PCR amplification. As illustrated in Figure 7.7, additional cycles can allow detectable levels of PCR products to be generated under conditions in which inhibition is substantial. The difficulty with this strategy is that increasing the cycle number will amplify noise as well as signal, thereby increasing the risk of false-positive results. Modest increases may be advantageous, but large increases in cycle number can raise the risk of spurious results.

Inhibitors can also compromise greatly the ability of real-time PCR to provide accurate quantification of the GMO content of foods and agricultural products. By convention, GMO content is reported as percent GMO. This is generally understood to signify the mass percent of any given species present in the product. Real-time quanti-

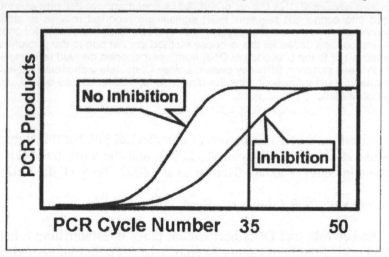

FIGURE 7.7. Effects of inhibitors on PCR reaction. Inhibitors slow the time course of the PCR reaction, but by increasing the cycle number, PCR reaction products can accumulate to substantial levels, allowing sensitive detection even in the presence of inhibitors. This strategy carries greater risk of generating false positive results, however.

tative PCR determines directly a molecular equivalent to mass percent, the number of copies of the GM target sequence present in the sample divided by the total number of genomes (for the species of interest) present in the sample. For example, for Roundup Ready soy, which contains one complete copy of the 35S promoter from cauliflower mosaic virus, one might measure the number of copies of the 35S promoter, and the number of copies of the soy lectin gene. Percent GMO would then be calculated using the formula:

$$\%GMO = \left(\frac{(\text{PCR signal for 35S promoter})}{(\text{PCR signal for soy lectin gene})} \right) \times 100$$

Because the GMO marker (35S promoter in the previous example) and the species marker (soy lectin gene) are both part of the same genome and of the same DNA preparation, any inhibitors of PCR that may be present will influence PCR amplification from these two DNA sequences equally, and, in principle, should not influence the calculated values for percent GMO. Although this logic can be used safely in cases where a moderate degree of inhibition is observed, it cannot be used in cases where inhibition becomes substantial. This limitation is illustrated in Figure 7.8, where we see that as the level of inhibitor increases and mean PCR signal decreases, the variation in PCR signal among replicate samples (represented by error bars on the bar graph) does not decrease proportionally to a decrease in the PCR signal itself. As a result, the precision falls off strongly as the degree of inhibition increases. For example, if percent GMO is 10 percent ± 15 percent for a sample lacking inhibitors, percent GMO might move to 10 percent ± 25 percent at 50 percent inhibition. In this case, precision is still sufficient to allow reasonable confidence in quantitative measurements. However, at 90 percent and 99 percent inhibition, variation among replicates might be 10 percent ± 50 percent and 10 percent ± 100 percent, respectively. In these cases, quantitative values become virtually meaningless because the magnitude of variability among replicates approaches that of the signal, itself. Thus, when inhibition becomes significant, this approach to compensating for inhibition fails. The alternative is to improve DNA purification procedures to remove inhibitory substances. This approach is more demanding, but it assures the ability to generate reliable quantitative data.

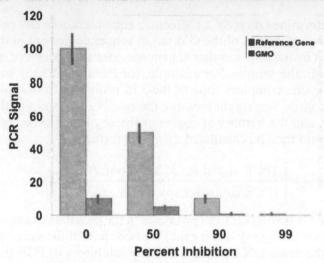

FIGURE 7.8. Inhibition increases variance of quantitative PCR analyses. The light bars indicate the PCR signal derived from a primer set targeted to a species-specific marker gene. The dark bars indicate the PCR signals from a primer set targeting a GMO-specific DNA sequence. The purpose of this figure is to illustrate that as inhibition increases, signal decreases in parallel for both the reference gene and GMO, but the size of the standard error of the mean (indicated by error bars) does not decrease proportionally. Thus, as the inhibition increases and signal decreases, the sizes of the error bars increase relative to the PCR signals. This results in increased variability in the calculated values of percent GMO as inhibition increases. Because the percent GMO values are rations, the variability in percent GMO values significantly increases as inhibition increases. Some labs have reported quantitative values, even when inhibition is 99 percent or more. Clearly, this is risky.

Dealing with DNA Degradation

Removal of inhibitors by improving DNA purification procedures is also useful in optimizing the quality of PCR data obtainable from sample types where partial degradation of DNA is unavoidable. Degradation is often a problem with samples derived from highly processed food products, in which the processing procedures actually cause significant DNA fragmentation. Referring to the earlier formula for percent GMO, it is clear that the value determined for a given sample should be independent of the degree of degradation of the sample DNA. This is because the GMO target sequence and the species-specific target sequence should be degraded in parallel since they are both part of the same genome. However, DNA degradation can influence significantly the precision of this quantity. As DNA degradation decreases the magnitude of the PCR signal, the magnitude of variation among replicate analyses does not decrease proportionally. Consequently, variation among replicate analyses will increase as the degree of DNA degradation increases in a manner similar to that discussed in the previous section for the case of PCR inhibitors.

This trend can be countered by increasing the amount of sample DNA added to the PCR reaction. However, if PCR inhibitors are present in the sample DNA preparation, the ability to add more DNA to the reaction will be limited. DNA purification methods have improved progressively in recent years. Thus, procedures are now available that can effectively remove inhibitors from most sample types. It is now possible to add larger amounts of DNA to PCR reactions, thereby improving the precision of analysis. With currently available DNA purification methods, 200 to 300 ng of DNA can routinely be added to 25 µl reactions, and for many food matrixes, up to 500 ng can be used. In such cases, the limiting factor in achieving satisfactory quantitative analytical results from a given sample becomes the total yield of DNA that can be obtained from a given sample.

A second, equally important approach that can be employed to reduce the impact of DNA degradation on sensitivity and precision of PCR analysis is to design primer sets whose recognition sites within the target DNA are as close together as possible. DNA degradation is a random cleavage process. The ability to generate PCR products (amplicons) from a given DNA target sequence is destroyed if a cleavage event occurs between the two primer binding sites that delimit the PCR target sequence. The probability that a cleavage event will occur between two primer recognition sites is directly proportional to the distance between those two sites, and thus falls off rapidly as amplicon size is reduced. For example, if degradative processes cleave the sample DNA randomly on average every 500 bp, the probability that a 100 bp amplification target will be cleaved is 100/500 or 20 percent. If, on the other hand, the amplification target is 300 bp, the probability of cleavage is 300/500 or 60 percent, and we can expect that 60 percent of target molecules will be degraded. Thus, by reducing amplicon size from 300 to 100 bp, the number of amplification targets is increased from 40 percent of total genomes present in the sample to 80 percent, which doubles the sensitivity. Therefore, we recommend that primers be designed to generate amplicons of 90 to 120 bp in length. In some cases even shorter target sequences can be used. The previous analysis also illustrates the importance of selecting GM and species-specific amplification target sequences that are close to the same size. If these targets are significantly different in size, degradation of sample DNA will significantly bias quantification.

PCR Amplification and Detection

The basic process of PCR amplification and detection was described earlier. PCR-based GMO detection methods can be placed into four categories, based on whether they are quantitative or qualitative, and on whether they employ primers that detect sequences unique to a single GMO or common to many different GMOs. The utility of these four categories of methods in answering specific analytical questions is summarized in Table 7.1 and discussed in more detail in subsequent sections. In all cases, PCR methods for GMO analysis must use standardized procedures that include carefully designed controls and carefully selected reference materials. These ensure that the method will achieve the following:

- The performance characteristics (at a minimum, sensitivity and specificity) of the analytical method must be verified in each analytical run.
- Consistency in analysis and interpretation of results must be ensured over time, from operator to operator, and from lab to lab.
- Malfunctions of analytical equipment and operator error in performance of the analysis should be reliably detected and corrected.
- Characteristics of the sample that create problems for the analysis or that interfere with achieving accurate results, such as partial degradation of the sample DNA or the presence of substances in the sample that inhibit PCR, should be detected.

Later sections of this chapter discuss in detail the controls, reference materials, and standard operating procedures required to ensure reliable results.

TABLE 7.1 Comparison of qualitative and quantitative methods

	Qualitative method	Quantitative method
Broad-spectrum primers	Broad-spectrum screening: Is genetically modified material present in the sample?	Rough quantification: Approximately how much genetically modified material is present in the sample?
Event-specific primers	Event-specific detection: Which GMO(s) is (are) present in the sample?	Precise quantification: How much of a particular GMO is present (one primer set)? How much total GMO is present (total measurements for all GMOs in the sample)?

Broad-Spectrum Screening Methods

The goal of GMO screening methods is to parsimoniously, but definitively, detect all transgenic varieties of interest. Such methods use "broad-spectrum" primer sets, that is, primer sets that target sequences present in many different GMOs. To design such a screening method, one examines the sequences present in the GMOs of interest and selects a series of primer sets capable of efficiently detecting all of those GMOs. The specific details and complexity of the method will depend on the sequence elements present in the GMOs of interest. If, among the many sequence elements present in that set of GMOs, a single element exists that is common to all of them, then a single primer set directed to that sequence element will be capable of detecting the GMOs.

In many cases, no single sequence element is common to all GMOs of interest. Therefore, it is necessary to design a series of primer sets that will provide definitive evidence for the presence or absence of each GMO of interest. For example, the cauliflower mosaic virus (CaMV) 35S promoter is present in all of the maize varieties commercialized to date except one GA21, a variety resistant to glyphosate. However, this GMO contains the terminator sequence from the nopaline synthetase gene from *Agrobacterium*. This sequence is also common to many, but not all genetically modified maize varieties. A PCR test that uses primers recognizing both the 35S promoter and the nopaline synthase (NOS) terminator should detect all commercialized transgenic maize varieties, since at least one of these primer sets will react with every transgenic maize event, and many events will react with both. If positive signals are detected with one or both of these primer sets, one can have reasonable confidence that at least one genetically modified maize variety is present. However, as will be discussed later, it is, in most cases, highly advisable to confirm this initial result using additional primer sets before reporting results.

Quantitative PCR analyses are often run using broad spectrum primer sets, such as the 35S promoter and the NOS terminator, for the purpose of obtaining a rough estimate of the GMO content of the sample. However, as is discussed in more detail in the following section, the results of such analyses are approximate because the sequence elements targeted by these broad-spectrum primer sets are present in different copy numbers in different GMO events or varieties.

Variety-Specific or Event-Specific GMO Detection Methods

Once GM material has been detected, it is often necessary to determine precisely which GMO or GMOs are present. This necessity arises because the importation of genetically modified foods or feed is contingent upon approval of these products for specific uses. National and regional differences in approval status for a given genetically modified crop can create substantial challenges for importation. Table 7.2 illustrates this situation using the example of maize. The two right-hand columns list all of the transgenic corn events or varieties that have been approved for cultivation in the United States. Of these, the top ten have actually been produced commercially on a large scale. On the right is the approval status of these products for human use in various international markets. Only four of the ten have been approved in the EU. Three have been approved in Norway and Switzerland, and nine in Japan.

Differences in approval status create challenges for exporters attempting to move maize or maize products into various markets. Not only is it necessary to quantify the total amount of genetically modified material present in the product in order to comply with labeling regulations in these countries, but it is also necessary to ensure that the product lot does not contain varieties or events that have not been approved for food use in the receiving country.

TABLE 7.2. National approvals of genetically modified maize

Event name	Brand name	Approved for human food			
		EU	Japan	Switz.	Norway
Syngenta E176	NatureGard/KnockOut	Yes	Yes	Yes	No
MON 810	YieldGard	Yes	Yes	Yes	Yes
Syngenta Bt11	Bt11	Yes	Yes	Yes	Yes
Aventis T25	Liberty Link	Yes	Yes	No	Yes
MON GA21	Roundup Ready	No	Yes	No	No
Aventis T14	Liberty Link	No	Yes	No	No
DeKalb DLL25	GR	No	Yes	No	No
DeKalb DBT-418	Bt Xtra	No	Yes	No	No
Aventis CBH-351	StarLink	No	No	No	No
MON NK603	Roundup Ready	No	Yes	No	No
Aventis (PGS) MS3	Not commercialized	No	No	No	No
Aventis MS6	Not commercialized	No	No	No	No
Pioneer 676/678/680	Not commercialized	No	No	No	No
MON 801	Not commercialized	No	No	No	No
MON 802	Not commercialized	No	No	No	No
MON 805	Not commercialized	No	No	No	No
MON 809	Not commercialized	No	No	No	No
MON 830/831/832	Not commercialized	No	No	No	No

The current status of labeling regulations in the European Union (EC, 1997, 1998) exemplifies the situation encountered in many parts of the globe. Products of unknown composition or products that contain greater than 1 percent genetically modified material must be labeled as "produced through gene modification." Products containing less than 1 percent genetically modified material do not require label, as long as the producer can provide (1) strong traceability documentation demonstrating that positive efforts were taken to avoid GMO admixture, and (2) test reports documenting that the GMO contents of the ingredients in the product comprise no more than 1 percent GMO. However, testing alone is not enough. The producer must have in place a strong and well-documented identity preservation (IP) or traceability system. In addition, it is necessary to demonstrate the complete absence of genetically modified maize events that have not been approved for food use in Europe.

To meet these needs, variety- or event-specific detection methods have been developed and are currently offered commercially for all transgenic soy, maize, and potato varieties. These include the following: RR soy, Novartis Bt176 maize, Novartis Bt11 maize, DeKalb DLL25 maize, DeKalb DB418 maize, Monsanto Mon 810 maize, Aventis T25 maize, Aventis T14 maize, Aventis CBH351 maize, Monsanto GA21 maize, Monsanto Newleaf potatoes, Monsanto Newleaf Plus potatoes, and Monsanto Newleaf Y potatoes. Whereas screening methods target DNA sequences that are common to many GMOs, varietal methods target sequences that are unique to the GMO of interest. This is accomplished by designing "bridging" primer sets. Such primer sets generate amplicons that span the junction between two transgenic sequence elements that have been joined in a manner unique to the GMO of interest. Figure 7.9 illustrates two classes of such sequences. The first class consists of internal junctions in which two sequence elements of the transgenic construct are joined in a unique manner (primer set B in Figure 7.9). Primer set B consists of one primer that recognizes sequence 1 and another that recognizes sequence 2. These two sequence elements do not exist adjacent to each other in nature, so the only case in which a PCR product will be made from this primer set will be when the DNA sample contains a transgenic construct in which sequence elements 1 and 2 are juxtaposed. Thus, this primer set is specific for any recombinant organism that carries the construct containing the junction of elements 1 and 2.

The second class consists of border junctions, where the transgenic construct is joined to genomic sequences of the host organism (primer set D in Figure 7.9). Primer set D includes one primer specific for sequence 5 and one specific for the maize genomic sequences flanking the recombinant gene. As with primer set B, PCR products will be made from primer set D only when the sample DNA contains the recombinant gene depicted in Figure 7.9, because only then will sequence 5 and the maize genomic sequences be juxtaposed.

Use of each of these two classes of bridging primer sets has advantages. Border junctions have the advantage of being truly event-specific because transgene insertion occurs randomly. Thus, different

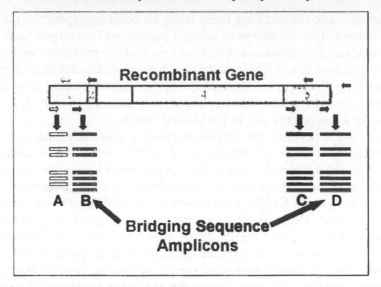

FIGURE 7.9. Bridging primers allow definitive detection of specific transgenic events. Primer sets B and D bridge junctions between two sequence elements fused during the construction of the recombinant gene. By selecting primer sets that bridge junctions that are unique to a given transgenic event, an assay can be developed that is highly specific for the detection of that single transgenic event. Primer set B bridges the junction between two internal elements in the trangenic construct, whereas primer set D bridges the junction between the transgenic construct and the host genome. The former will detect any GMO that was produced using the recombinant gene construct shown in the illustration. The latter will detect only the single transformation event shown because the insertion site for the construct is random and will be different for each transformation event.

transgenic events derived from the same transgenic construct will possess different flanking sequences. This property allows one to differentiate individual events where necessary. For example, in the case of Liberty Link soy, seven different events have been approved for commercial production in the United States. Use of primers specific for the border junctions of these events could differentiate among them. On the other hand, since the gene constructs inserted into all of these events were the same or highly similar, a single primer set that targets an internal junction sequence unique to that particular construct would detect all of the Liberty Link soy events that have been approved, and differentiate them from all other transgenic crops on the market. The advantage in using a primer set that targets such an internal junction sequence is that, in cases where multiple events are in the marketplace, a single primer set can be used to detect all events containing the construct of interest. This is useful, however, only if all events are treated as a class by regulatory authorities, which is the case for Liberty Link soy in the United States.

It is worth pointing out that event-/variety-specific primer sets are essential to accurate quantification of GMO content. Until recently, the typical approach to quantification has been to use one or more broad-spectrum primer sets that recognize common transgenic elements, such as the CaMV 35S promoter, the NOS terminator, and the Cry1Ab gene. These elements are present in different copy numbers in different transgenic events, as is shown in Table 7.3. Therefore, they cannot be used for accurate quantification of percent GMO in samples in which more than one event is, or may be, present. Because complex mixtures of events are not the exception but the rule for real-world samples, such broad-spectrum primer sets seldom provide accurate quantification. In contrast, event-specific primers can be used to achieve accurate quantification of each individual variety or event, and total GMO content can be calculated by summing the values for the individual events.

TABLE 7.3. Copy number of common sequence elements

Event name	35S-P	35S-T	NOS-T
Nov/Myc E176	2	2	0
AgrEvo T14	3	3	0
AgrEvo T25	1	1	0
AgrEvo CBH-351	4	1	4
DeKalb DBT-418	3	0	0
DeKalb DLL25	1	0	0
MON 810	1	0	1
MON GA21	0	0	2
NK Bt11	2	0	2

GMO Assay Design and Quality Control

Whether qualitative or quantitative, broad-spectrum or event-specific, the core of every GMO assay is a set of PCR reactions containing primers specific for the GMO(s) of interest. However, the results obtained from such reactions provide insufficient information by themselves to (1) conclude with confidence whether genetically modified material is actually present in the sample; (2) quantify, if necessary, the level of genetically modified material present; and (3) understand the limitations of the analysis of a particular sample. To fulfill these needs, additional assay design elements, including additional primers, are required. These elements enable the analyst to achieve the following specific objectives:

1. Detect the presence of inhibitors that would interfere with the PCR process.
2. Assess whether the sample DNA is degraded.
3. Verify that the PCR reagents and equipment are functioning properly.
4. Determine the level of sensitivity (limit of detection) of the PCR method and verify that the PCR process is operating to a consistent level of sensitivity from run to run.
5. Confirm positive and negative results.

In addition to carefully thought-out design elements, an analytical method of practical utility must be embedded in the context of a comprehensive quality assurance/quality control system. Only then can the method achieve the degree of reliability and consistency that regulators and the industry require in order to effectively guide their work. The following sections discuss the required assay design features and how they can be used within a QA/QC system to effect reliable and practical GMO analyses.

Conventional or Endpoint PCR-Assay Design and Quality Control

A number of methods have been reported for broad spectrum screening of GMOs (Brodmann et al., 1997; Jaccaud et al., 2003; Lipp et al., 1999; Pietsch et al., 1997). Table 7.4 outlines the design

TABLE 7.4 Design features of PCR methods

	PCR reagent control— No DNA	Internal control DNA template	Non-GMO reference DNA	GMO reference DNA	Sample DNA preparation	Sample DNA preparation plus internal control DNA template
Internal control primer set	Rxn 1 & 2	Rxn 3 & 4				Rxn 5 & 6
Species-specific reference gene primer set	Rxn 7 & 8			Rxn 9 & 10	Rxn 11 & 12	
GMO primer set	Rxn 13 & 14		Rxn 15 & 16	Rxn 17 & 18	Rxn 19 & 20	

features of such endpoint PCR methods. Every GMO analysis consists of at least three reaction series, each of which serves a specific function in the analytical procedure. Reactions 1 through 6 constitute the internal control reaction series. The purpose of this series is to assess whether PCR inhibitors are present in the DNA extracted from the unknown sample, and to serve as an internal reference point to verify that the sensitivity of the PCR system is consistent from run to run. This is accomplished by comparing the efficiency with which the control DNA template is amplified in the absence of sample DNA (reactions 3 and 4) or in the presence of sample DNA (reactions 5 and 6). Reactions 1 and 2 are a reagent control to ensure that the generation of the internal control amplicon is dependent on the addition of an internal control template. Figure 7.10 illustrates the possible results that can be obtained from the internal control series of reactions. For any given PCR run, reactions 1 through 4 are run once, whereas reactions 5 and 6 are run for every sample in the PCR run.

The internal control DNA template is a DNA molecule of known sequence that is normally not present in food products. The key factor in designing the internal control series of reactions is to establish the amount of internal control DNA template that should be added to each reaction. This template should be added at a level that is low enough to ensure that the PCR band intensity will be responsive to inhibitors, if present. At the same time, the level should be high enough to ensure ready and consistent detection of the signal in the absence of inhibitors. This level must be determined empirically, since efficiency of amplification varies from template to template.

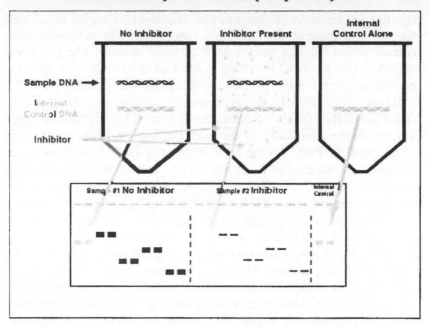

FIGURE 7.10. The internal control—A control for the presence of PCR inhibitors. A defined concentration of a known target DNA molecule (internal control template) can be added to the sample (left-hand and central tubes), and amplification can be compared to amplification from that target alone (right-hand tube). If, as in the case of the central tube, inhibitory substances are present in the sample DNA preparation, amplification from the internal control DNA template will be reduced. In contrast, if no inhibitory substances are present, as in the case of the left-hand tube, amplification from the internal control template will generate a PCR signal corresponding in intensity to that observed when the internal control template is amplified by itself.

In conjunction with the internal control reaction series, the species-specific reference gene reaction series generates additional information regarding the quality of the sample DNA. This series employs a primer set that targets a gene that is endogenous to the species relevant to any given sample. The species-specific reference gene must have two key characteristics. First, it should be constant in copy number, a feature typical of nuclear genes. Second, it should be present in both traditional and transgenic lines of the relevant species. If a given sample is expected to contain only soybeans and their derivatives, then only one species-specific reference gene reaction series will be required for analysis of that sample. However, if maize or oilseed rape are also present, then separate species-specific reference gene reaction series must be run for each relevant species. The key comparison in this series of reactions is the extent to which the species-specific reference gene primer set generates amplicons from the GMO reference DNA (reactions 9 and 10), versus the extent of amplification of the sample DNA (reactions 11 and 12). For this comparison, either the non-GMO or the GMO reference DNA preparation can be used. In Table 7.4 the GMO reference DNA (reactions 9 and 10) was used. Reactions 7 and 8 are a reagent blank, demonstrating that generation of the species-specific reference gene amplicon is dependent on the presence of DNA derived from the species of interest.

The results emerging from the species-specific reference gene reaction series is a combined measure of the presence of inhibitors in the sample DNA preparation and the occurrence of degradation of the sample DNA preparation. By comparing these results with those obtained from the internal control reaction series, one can resolve these two variables and obtain a reasonably accurate estimate of both inhibition and degradation. Figure 7.11 illustrates the effect of degradation on PCR amplification of the reference gene. These reactions also provide useful information regarding the general performance of PCR reagents and equipment, in that, if defective reagents or equipment are used, amplification of the species-specific reference gene and the internal control template will be anomalous.

Results of the positive control and the internal control reaction series must be interpreted together. If these results indicate that inhibitors are present, then it is advisable, as discussed previously, to repeat the isolation of DNA from that sample and perform another PCR analysis. This is critical, because the sensitivity of the GMO analysis is reduced in the presence of inhibitors. Similarly, partial degradation of the sample DNA can significantly reduce sensitivity of the analysis. In cases where the partially degraded sample appears not to contain PCR inhibitors, sensitivity can be increased by adding more sample DNA to repeat PCR reactions. However, if inhibitory substances are present, these must be removed by improving the DNA extraction process before DNA input can be increased.

The GMO primer set reaction series provides information regarding the presence of a specific GMO or set of GMOs in the sample. In this series, reactions 17 through 20 are key because they provide

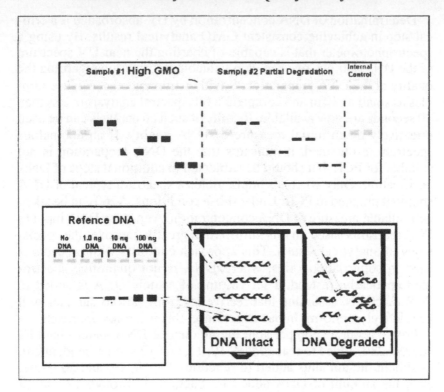

FIGURE 7.11. Positive control using primers specific for a species-specific reference gene—A control for DNA degradation. A primer set targeting a gene common to all varieties (conventional or trangenic) of the crop species of interest can be used to assess the integrity of the sample DNA. DNA degredation will reduce the signal generated from this primer set. The signal is detected by comparing the intensity of PCR signals generated with this primer set from reference DNA from the species and sample. A reduction in signal with sample DNA, as seen in sample #2, indicates partial degradation, assuming that the internal control indicates absence of inhibitory substances.

information regarding the GMO content of the sample DNA preparation (reactions 19 and 20) compared to a bona fide sample of the GMO DNA of interest (reactions 17 and 18). Reactions 13 through 16 are negative controls, demonstrating that generation of the amplicon typical of the GMO of interest is dependent upon the presence of DNA derived from the GMO.

Determination of DNA concentration by UV absorbance is a critical step in achieving consistent GMO analytical results. By using a spectrophotometer that is capable of charting the near-UV spectrum of the DNA sample, one can gain valuable information regarding the quality of the DNA preparation. Small instruments that can use samples as small as 10 µl and complete a full spectral analysis in less than 30 seconds are now available. Results of such an analysis can be used effectively as an initial measure of DNA quality. If an anomalous spectrum is obtained, it indicates that the DNA preparation is not suitable for PCR, but should be subjected to additional steps of DNA purification. Only when a sample yields a spectrum typical of DNA should it proceed to PCR. Under those conditions, A260 can be taken as a reliable measure of DNA concentration. Adjusting DNA input to PCR reactions based on this information greatly improves the consistency of analytical results. This approach contrasts with procedures used in some laboratories, where DNA is not quantified spectrophotometically. Instead, a set volume of sample DNA is added to PCR reactions, assuming that recoveries from various samples will be relatively uniform. In our experience, this is a risky approach.

For an accurate comparison, the amount of DNA, determined by A260/A280, added to reactions 9, 10, 11, and 12 should be identical, as should the amounts added to reactions 17, 18, 19, and 20. However, the amount of DNA added to reactions 9 through 12 and 17 through 20 will differ significantly. The amount of DNA present in the GMO reference DNA reaction series (reactions 17 and 18) is adjusted to be equal to the limit of detection of the PCR method. Typically, this is in the range of 0.01 percent GMO. The intensity of signal observed from reactions 17 and 18 provides an indication of whether the PCR system is functioning to the expected sensitivity. If it is not functioning as expected, then the analysis is repeated with different reagent lots. An amount of sample DNA is added to reactions 19 and 20 that is equal to the amount of GMO reference DNA added to reactions 17 and 18. Comparing the intensities of bands in these two sets of reactions provides a rough measure of the GMO content of the sample. However, since this is endpoint PCR, such a comparison is not a particularly accurate measure of GMO content. Nevertheless, useful information is often obtained.

If the limit of GMO detection is 0.01 percent, and the concentration of DNA added to reactions 9 through 12 is adjusted to operate in roughly the same sensitivity range as reactions 17 through 20, then the amount of DNA added to these reactions should be approximately 1/1,000th of that in reactions 17 through 20. This is because approximately 1,000 conventional genomes are present for every genetically modified genome in a sample containing 0.1 percent GMO. Thus, if 1/1,000th the DNA input is used, this will ensure that the PCR signal observed with reactions 9 through 12 will be in a range that is sensitive to the presence of degradation.

The previous example is for a GMO screening test that targets a single GMO-specific sequence element. In assays where additional transgenic sequences are targeted, separate series of reactions analogous to reaction series 13 through 20 will be carried out. The nature of the GMOs of interest will determine the genetic sequences targeted in the assay. If targets are selected that are unique to an individual GMO, then the test will detect that GMO only. On the other hand, if the test targets sequences common to many different GMOs, then the test will serve as a screen for all of those GMOs. As discussed earlier, both of these kinds of tests have important functions.

Semiquantitative Approaches to PCR Analysis

There are both endpoint and kinetic approaches to semiquantification of GMO content. QC-PCR (quantitative-competitive-PCR) is the most reliable and commonly used approach (Hübner, Suder, and Lüthy, 1999a,b; Lüthy, 1999; Pietsch et al., 1999; Wurz et al., 1999). In this approach, a synthetic template is added to the PCR reaction in addition to sample DNA. This template contains the same primer-annealing sites as the GMO target sequence of interest, but has been modified to contain a few extra or a few fewer bases so that amplicons generated from this template will be longer or shorter than the GMO amplicon and can be distinguished from it electrophoretically. During the process of bringing the amplification reaction to endpoint, competition takes place between the bona fide GMO template and the competitor template. Because certain reagents, in particular Taq polymerase, primers, and dXTPs, become limiting during the PCR reaction, any difference that exists initially between the concentrations of the GMO template and the competitor template will be accentuated in the final amplification products. The more abundant of the two templates will compete more effectively and be amplified more frequently than the template of lesser abundance. This will not occur only when the initial abundance of the two templates is very close to equal. Under those conditions, the two templates will compete with equal effectiveness for amplification reagents, and will, therefore, be amplified equally. By carrying out many reactions, each containing a different concentration of competitor, one can find the concentration of competitor template that is equal to the concentration of the GMO target template, and thus determine the GMO content of the sample with relative accuracy. This same procedure is then carried out to determine the concentration of the species-specific reference gene, and the percent GMO is calculated from these two values using the following formula:

$$\%GMO = \left(\frac{\text{Concentration of GMO Target Sequence}}{\text{Concentration of Species-Specific Reference Gene}} \right) \times 100$$

In principle, this procedure can be highly quantitative (Hübner, Studer, and Lüthy, 1999a,b; Hübner, et al., 1999c; Lüthy, 1999; Pietsch et al., 1999). However, because one must carry out many PCR reactions to achieve this, the method is more often used in a semiquantitative format in which a single concentration of competitor is used, defining a threshold that is the basis against which the amount of transgenic material in the sample is judged to be greater or less (Wurz et al., 1999).

The second approach to semiquantification is a freeze-frame approach (Tozzini et al., 2000). As illustrated in Figure 7.12, if one can adjust PCR conditions such that the reactions are terminated at a time (designated by line A) when amplification of most of the samples in the run has not plateaued, then band intensity will be roughly proportional to GMO content, and one can, therefore, use band intensity to assess GMO content relative to reference reactions containing known amounts of genetically modified material. The weakness of this method is that it works well only with samples that are highly uniform in DNA quality and free from DNA degradation and inhibitors, where direct comparison can be made with reference reactions. Clearly, the controls that detect the presence of PCR inhibitors and DNA degradation are critical in such an assay. Quantification is justified only if strong evidence exists that neither inhibition nor degradation has occurred. In addition, this approach provides order of magnitude accuracy in quantification only.

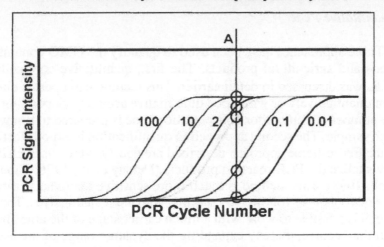

FIGURE 7.12. PCR reaction profiles. Typical reaction profiles for PCR, indicating a point, marked by line A, where all reactions are terminated and subjected to electrophoretic analysis. If a series of reactions has not reached a plateau, the level of PCR products indicated by circles on line A will be roughly proportional to the GMO content of the original samples.

The same controls described previously for qualitative screening methods are required in semiquantitative configurations of PCR. However, the process requires a more extensive set of external standards that can serve as the basis for generating a calibration curve for estimating the GMO content of the sample. Typically, the calibration curve will contain concentrations ranging from 0.01 percent to 2 percent GMO. As in the case of qualitative assays, these external controls also provide an independent point of reference for verifying the sensitivity of the assay.

It should be pointed out that verifying that the assay is operating to the intended limit of detection (LOD) using external standards does not demonstrate that the assay is operating to that same LOD for any given sample. The external controls simply verify that, under conditions in which the sample DNA is free from degradation and inhibitors, the observed LOD can be achieved. The other controls—in particular the internal control—provide information on the individual sample that allows one to conclude whether the LOD obtained with reference materials accurately reflects the LOD for the sample (Jankiewicz et al., 1999).

Quantitative PCR

Three approaches have been used to quantify the GMO content of foods and agricultural products. The first, quantitative-competitive PCR, was discussed in detail earlier. This method can operate either semiquantitatively or with good quantitative accuracy, depending on the number of competitor concentrations one is prepared to run with each sample. The second approach to quantification is a modification of the freeze-frame approach described previously, which uses ELISA to visualize the PCR reaction products (Demay et al., 1996; Taoufik et al., 1998). This method is much more sensitive than visualization by electrophoresis followed by staining with intercalating dyes. Therefore, it is possible to work with a much earlier stage of the amplification time course, thereby expanding the dynamic range of the assay and improving quantitative accuracy.

The third approach to quantification, real-time PCR (RT-PCR), was first developed in 1992 (Higuchi et al., 1992). Real-time PCR has found broad application in GMO analysis (e.g., see Brodmann et al., 2002; Shindo et al., 2002; Vaitilingom et al., 1999). The attraction of this methodology is the relative ease with which quantitative data can be generated compared to other methods. In RT-PCR, the PCR reaction is linked to a second process that generates a single fluorescent reporter molecule for every amplicon that is generated during PCR. Using these methods, the full-time course of as many as 96 PCR reactions can be charted in detail simultaneously. Figure 7.13 presents one such time course, recorded using a Bio-Rad iCycler iQ system. A straight-line plot, as shown in Figure 7.14, is obtained when one plots the number of cycles required for a set of samples containing known amounts of GM material to achieve a certain threshold of fluorescence, indicated by the horizontal line in the figure (line A), versus the GMO content of the samples (10, 1, 0.1, 0.01, and 0 percent GM soy). The principles presented in Figures 7.13 and 7.14 can be used to quantify accurately the GMO content of food and agricultural products. Each series of analyses includes analysis of a full set of standards, giving rise to a calibration curve similar to that shown in Figure 7.14. The results obtained for individual unknown samples are compared to the calibration curve to determine the GMO content of those unknowns. The real-time instrumentation systems, such as ABI TaqMan and Bio-Rad iCycler iQ, automate this analytical procedure.

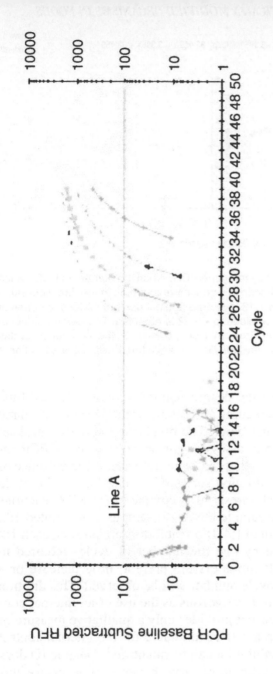

FIGURE 7.13. Real-time quanitative PCR analysis. Fluorescence profiles generated during real-time PCR, plotted as log fluorescence signal (arbitrary units) versus PCR cycle number. The number of cycles required to generate the threshold level of fluorescence indicated by line A is proportional to the target concentration in the sample (plotted in Figure 7.14). In principle, the threshold specified by line A can be set at any point within the logarithmic portion of the fluorescence profile.

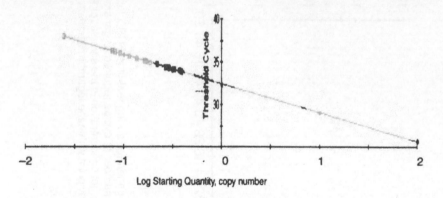

FIGURE 7.14. Quantification by real-time PCR. When the number of PCR cycles required to generate a fluorescence signal corresponding to the threshold value specified by line A in Figure 7.13 is plotted against percent GMO of a series of standards, a line with a slope close to −3.32 is obtained. The percent GMO of any sample can be deduced from this plot, based on the number of cycles required for the sample to achieve the fluorescence threshold indicated by line A in Figure 7.13.

Table 7.5 summarizes the design features essential to real-time PCR analysis. As with screening and semiquantitative analysis, three series of reactions are carried out. The first is the internal control series, which is performed to assess the presence/absence of PCR inhibitors. Another approach that can be used to assess the presence of inhibitors is to set up a dilution series of sample DNA. The cycle number required to reach threshold is compared for PCR reactions containing (1) undiluted sample DNA, (2) sample DNA diluted 1:2, and (3) sample DNA diluted 1:4. If no inhibitors are present, each 1:2 dilution should increase by one the number of cycles required to achieve the threshold. If inhibitors are present, little increase or a nonintegral increase in cycle number will be observed. This dilution approach is not considered as rigorous as the use of an internal control reaction series, because it provides only a qualitative measure of inhibition. By setting up a calibration curve for the internal control template, the extent of inhibition can be quantified. Genetic ID does not routinely run such a calibration curve, however, if necessary, this

TABLE 7.5. Design features of real-time PCR analysis

	PCR reagent control— No DNA	Internal control DNA template	Non-GMO reference DNA	GMO reference DNA calibration curve	Sample DNA preparation (duplicates of two DNA extractions)	Sample DNA preparation plus internal control DNA template (singlets of each DNA extraction)
Internal control primer set	Rxn 1 & 2	Rxn 3 & 4				Rxn 5 & 6
Species-specific reference gene primer set	Rxn 7 & 8			Rxn 9 through 18	Rxn 19 through 22	
GMO primer set	Rxn 23 & 24			Rxn 25 through 34	Rxn 35 through 38	

can be done for critical samples. In fact, inhibition is not a large problem in general, since highly efficient DNA purification procedures have been developed that remove inhibitors from virtually all samples, including highly complex multi-ingredient products. The greater limitation to reliable quantification of such samples is recovery of sufficient PCR-active DNA to enable quantification in cases where sample DNA is partially degraded. Degradation problems are obvious when the internal control reaction series indicates absence of inhibition, yet the signal from the species-specific reference gene primer set is lower than would be expected based on the DNA input to those reactions. Scaling up purification can improve recovery by five- to 20-fold. However, for samples of certain types, such as some grades of highly processed starch, recovery can still be insufficient to obtain enough DNA for a full quantitative analysis, including all necessary controls. With such materials, recoveries are highly dependent on the batch of material and cannot be predicted before analysis is carried out. In cases in which recoveries are not sufficient for complete analysis, results should be reported in qualitative format.

To quantify the species-specific reference gene in the sample, reactions 19 through 22 are compared with the calibration curve consisting of reactions 9 through 18. Likewise, to quantify the GMO target sequence, reactions 35 through 38 are compared with the calibration curve consisting of reactions 25 through 34. Percent GMO is then calculated from these two values, using the formula presented earlier. These two series of reactions can be run in separate reaction tubes. However, it has been found that multiplexing them actually improves precision (Vaitilingom et al., 1999). Since most of the real-time PCR equipment offered today has the capacity to monitor multiple wavelengths simultaneously, multiplexing is highly recommended.

Assay Design Features to Confirm Results

Duplicate Analyses

An essential confirmatory measure in all analyses is that the analysis must be carried out in duplicate. The duplicate analyses should be carried out independently, beginning with representative subsamples taken from the submitted field sample. DNA should be extracted independently from these subsamples, followed by separate PCR analyses. The results must agree or the whole analytical process must be repeated. Independent analysis of duplicate samples is an effective way to catch and correct a wide range of "operator errors," the mistakes that inevitably occur when even the most conscientious and well-trained technical staff carry out complex procedures, such as those required for GMO analysis. The principle behind the use of duplicate analyses is that it is unlikely that the operator will make the same mistake with both independently processed subsamples. Thus one duplicate serves as a comparator for the other, and if they do not yield similar results, it is concluded that a mistake may have been made in analysis of one of the two samples and the whole analysis must be repeated. The repeat analysis is again carried out in duplicate, and only when comparable results are obtained with duplicate analyses can results be reported with confidence. With real-time PCR, it is essential to conduct duplicates at the PCR level, as well. Thus, for real-time quantitative analyses, at least four replicates are run for each sample submitted to the laboratory.

Additional Confirmatory Analyses

To further increase confidence in analytical results, additional confirmatory measures can be implemented. According to Swiss and German law, putative positive analytical results must be confirmed by further analysis to verify that the amplicons correspond to the expected sequence (German Federal Foodstuff Act, 1998; *Schweizerisches Lebensmitlebuch,* 1998). Two approaches are recommended, Southern hybridization of PCR products to a probe known to be homologous to the bona fide amplicon of interest, or cleavage of the PCR products into fragments of expected size using restriction endonucleases. A third method that accomplishes the same objective is inherent in real-time PCR with Taqman, Molecular Beacon, and FRET probes. In each case, the appearance of a real-time signal is dependent on homology between the generated amplicons and the real-time probe. Thus successful generation of a fluorescence signal inherently embodies a confirmation of the identity of the amplicon. All of these methods are effective in identifying the amplicons generated during the PCR reaction. However, they fail to differentiate between PCR fragments generated from a sample containing bona fide transgenic sequences and a sample accidentally contaminated with amplicons corresponding to that sequence. More definitive confirmation can be obtained by carrying out additional independent PCR amplifications with primer sets that target a second site that will be present if, and only if, the site targeted by the first PCR amplification is present. If such primer sets are used, not in a second PCR run, but as a routine part of the initial analysis, two advantages are gained. First, analytical throughput is accelerated, since definitive, confirmed results are obtained in a single PCR run. Second, the use of multiple primer sets provides greater certainty in avoiding false negatives as well as false positives.

Developing and Implementing GMO Testing Methods

We have now considered the basic principles of GMO analysis by PCR. In the following section, we consider the process that is required to develop, implement, validate, and standardize a GMO testing method. The points presented here are based upon Genetic ID's years of experience in method design, development, and validation. They are also consistent with work in progress through the European Committee for Standardization (CEN/TC275/WG11) and the International Organization for Standardization (ISO/TC34/WG7), which are working together to develop harmonized, standard methods for GMO testing. The first three documents to emerge from this development process were published for parallel enquiry at CEN and ISO in

November 2002 (CEN/ISO, 2002a,b,c). Validation procedures have also been discussed at length in the literature (Anklam, 1999; Anklam, Gadani, et al., 2002; Anklam, Heinze, et al, 2002; Bertheau et al., 2002).

Two approaches are used to establish GMO testing capacity in the laboratory. Methods can be developed de novo, or they can be transferred into the laboratory from the published literature. We will discuss both approaches.

Developing GMO Testing Methods De Novo

The prerequisites for de novo method development are two: First, information must be available on the DNA sequence of the genetic target of the contemplated PCR method. Such information can be obtained usually by integrating information gleaned from documents submitted to regulatory agencies by the developer of the particular GMO of interest, and from gene-sequence databases. The second requirement for method development is the availability of a verified sample of the GMO of interest. This is essential, because verified material is required for the empirical assessment, optimization, standardization, and validation procedures every method must undergo.

At present, the Institute for Reference Materials and Measurements (IRMM) of the European Commission Joint Research Centre is the only organization offering GMO reference materials prepared according to the appropriate ISO guidelines (Trapmann et al., 2002). The IRMM markets these reference materials through Fluka, a chemical supply house. A limited selection of reference materials is available currently. However, when the revised EU regulation on labeling of GMOs comes into force, reference materials are expected to become more widely available because developers will be required to provide reference materials to the authorities if they wish their products to be approved for use in Europe. It may, however, still be difficult to gain access to reference materials for GMOs not approved in Europe.

Primer Design from Target Sequence Data

The first step in method development is to integrate all relevant sequence information related to the GMO of interest. Based on this information, several primer sets are designed that, in principle, should detect the GMO of interest. Typically, 8 to 12 primer sets are prepared for initial evaluation. The second step is an initial empirical assessment of the performance of the candidate primer sets. This leads to identification of two or three prime candidate sets, which are then subjected to greater scrutiny. The reaction conditions used for the preliminary screen are normally the standard reaction conditions most commonly used in the laboratory. In subsequent optimization work, we recommend adjusting concentrations of various reaction components, such as Mg^{++}, salt, and other additives. It is not recommended, however, to alter the temperature profile of the thermocycler, because it is highly desirable for all PCR systems used in the laboratory to use the same thermocycler conditions. This greatly simplifies and streamlines the work flow in the laboratory, making it possible to place reactions for several different target sequences and several different GMOs, in a single thermocycler run. In contrast, if different conditions are required, multiple, time-consuming thermocycler runs will be required for the PCR analysis of different gene targets.

Initial primer evaluation should assess the sensitivity of the amplification process, since some targets are amplified much more efficiently than others, and therefore yield more sensitive tests. It is also critical to assess specificity, since low specificity can lead to generation of false-positive results. Likewise, some primer sets will generate, not a single amplification product, but additional artifactual bands that can confuse interpretation of results. Finally, results obtained for replicate analyses should be consistent in their intensity. Primer sets displaying low specificity, extraneous bands, or variability between replicates should be eliminated from further evaluation.

Standardization for Routine Analytical Use

Once basic optimization has been accomplished using highly purified reference preparations of target DNA, the method must be tested with DNA from various sample types prepared by the methods designed for rapid sample throughput in the routine analytical laboratory. In general, little difference in performance will be observed with such DNA preparations, since those methods should already be thoroughly optimized. However, it is still prudent to conduct an evaluation.

Technology Transfer from the R&D Lab to the Routine Analytical Laboratory Environment

Initial assessment of the method in the analytical laboratory will examine the following parameters: sensitivity, specificity, accuracy, precision, reproducibility, robustness, operational utility and convenience, and economic practicality. This evaluation will involve the following:

1. Assessment in the hands of the same technician within the same analytical run.
2. Assessment in the hands of different technicians on different days.
3. Verification that results are consistent with different reagent lots.
4. Assessment of all sample types to which the method is applicable.

Based on the outcome of the initial assessment in the analytical laboratory, the method is taken through additional cycles of refinement until it is optimized. Once optimization is achieved, a final standard operating procedure is prepared, reviewed, and provided to the analytical laboratory for beta testing, using routine samples. It is preferred that the beta testing is done in parallel with analysis using an established method. If not, analysis must be conducted with additional controls, standard reference materials, and other safeguards to ensure that substandard performance will be detected and sufficient information generated to allow the operator to identify, understand, and correct the causes of inadequate performance.

External Validation by Laboratory Accreditor

Once a new method is deemed to be operating satisfactorily in the analytical laboratory, it should undergo validation by the independent accreditation body responsible for the laboratory's accreditation program. This will take place in the context of the preexisting, comprehensive accreditation program of the laboratory. To achieve validation, the laboratory conducts a series of studies designed to establish and document the performance of the method for commercial use. For this purpose, data from the beta-testing studies, described previously, can be used, along with any other studies necessary to meet the requirements of the accrediting body.

Transferring Methods from the Literature to the Laboratory

The basic steps required to integrate a published method into the services offered by an analytical laboratory are essentially the same as those outlined earlier for de novo method development, except that the primer sequences will be available in the publication, along with approximate conditions for PCR reactions. Whether a method is developed de novo or transferred from the literature, it is still essential to conduct an in-depth empirical assessment and optimization of the method. To achieve this, it is necessary to have access to reference material for the specific GMO(s) that the method targets. Likewise, all of the steps of adapting and standardizing the method for routine analytical use must be undertaken to ensure that the method will be operated in a consistent and reliable manner. Finally, the method must undergo accreditation as described previously. In our experience, the detail and accuracy of published methods are variable. In most cases, the published methods provide useful primer sequence information, but most of the other work required for de novo method development must still be done for adaptation of methods from the literature.

Maintaining Uniformly High Standards of Performance in the GMO Testing Laboratory

The following section is based on our experience in maintaining the quality of GMO analytical services within individual labs, maintaining uniformly high quality testing performance within the three laboratories that Genetic ID owns and operates in the United States, Germany, and Japan, and maintaining adequate consistency of analytical quality within the Global Laboratory Alliance, a group of 16 private and government laboratories, distributed around the world, which have licensed Genetic ID's GMO testing methods, and which are expected to operate these methods to a uniform set of standards and operating procedures.

Standard Operating Procedures

To ensure consistency and quality of GMO testing, a thorough and comprehensive system of standard operating procedures (SOPs) is essential. This system must include orderly and regular procedures for incorporating new methods into the system and for modifying existing methods as needed. SOPs are required for all laboratory manipulations, such as sample processing, DNA extraction, carrying out PCR reactions, and analyzing PCR products by electrophoresis or other methods. However, it is equally important to have standardized procedures for interpreting and reporting results. Standardized analytical methods are of little use if every individual who uses those methods interprets the results differently.

Laboratory Performance Assessment

Performance assessment programs are required to ensure that all procedures are being performed correctly and accurately. Such programs typically operate on at least two levels. First, laboratories operate an internal program in which they run "check samples" at a frequency proportional to the number of commercial samples analyzed by the laboratory. For example, one check sample might be analyzed for every ten "real" samples. The quality manager should formulate these samples and introduce them into the analytical stream labeled as any other sample, such that the technicians do not know which samples are real and which are not. Results of the check sample program should be reviewed on a regular basis with technicians and with management to keep quality specifications on track. It is important that check samples are representative of the range of sample types commonly analyzed by the laboratory. This ensures that all sample types are analyzed with equal care by technicians, and that the effectiveness of all protocols is assessed on a regular basis.

Second, laboratories should participate in external performance assessment schemes on a frequent basis. Several organizations offer such programs, including the U.S. Department of Agriculture, the U.K. Food Analysis Performance Assessment Scheme, the American Oil Chemists' Society, and the American Association of Cereal Chemists.

Laboratory Accreditation

Regular evaluation of analytical laboratories by an independent, third-party accrediting body is essential to verify the quality of analytical services provided. This evaluation must be conducted to a well-established, rigorous standard. The most widely accepted standard for analytical laboratories is ISO Guide 17025 (ISO, 1999). Accreditation to this standard is based on the following:

1. Inspection of laboratories on a yearly basis
2. Evaluation of credentials of technical and scientific staff
3. Evaluation of quality manual for the analytical operation
4. Evaluation of extensive validation data for each analytical method contained within the scope of accreditation

The scope of accreditation is for specific methods in the context of a specific laboratory environment. Thus, if a method is in use in multiple laboratories, it must be validated independently in each. That is, each laboratory must provide its own validation data. The utility of accreditation is highly dependent on the reputation and level of acceptance of the accreditation body. For example, reciprocity has not been established between U.S. and EU accrediting bodies. Thus, if a laboratory wishes to provide testing services in Europe, it is prudent to undergo accreditation by an EU-recognized accreditation body.

International Standardization of GMO Testing Methods

Laws requiring the labeling of foods consisting of GMOs, or containing ingredients derived from them, make necessary the availability of standardized methods for analysis of GMOs. Not only is GMO testing required, but quantitative methods are required, as well, because many of the national laws specify quantitative thresholds that trigger labeling of foods as genetically modified. The development of standardized methods for GMO testing has lagged behind the introduction of GMOs into the food system, as well as the enactment of labeling laws. At present, Japan (JMHLW, 2002a,b), New Zealand (NZMAF, 2002), Germany (German Federal Foodstuffs Act, 1998), and Switzerland (*Schweizerisches Lebensmittlebuch,* 1998) have all established official testing methods for some GMOs. However, coverage is not comprehensive. Because of the global nature of the food production system, the methods standardization process must be global in scope. Importers and exporters need to have confidence that the tests that they conduct in different areas of the world are consistent in sensitivity and specificity, and are of similar accuracy and precision. Only then can they be confident that the products released for shipment from one port, based on test results, will be found acceptable when they reach a port on the other side of the globe.

Several initiatives are in motion to develop, standardize, and validate methods. Most prominent are the CEN and ISO efforts mentioned earlier. Representatives from many countries are participating in this process. On a more localized level, a network of official government reference laboratories for GMO testing, the European Network of GMO Laboratories (ENGL), was established by the Joint Research Centre of the European Council (JRC-EC, 2002). This network encompasses nearly 50 official control laboratories, each appointed by the national authority of the corresponding EU member state. One of the network's objectives is to develop and standardize testing methods that will respond to testing needs evolving out of the EU's legislation on GMO labeling. Initiatives are also in progress in the seed, tobacco, and cereals industries.

One important point to note is that the initiatives mentioned previously will require years to achieve completion. In the meantime, the food and agricultural industries must find strategies for ensuring consistent compliance with labeling laws. One initiative that has provided an answer to this need is the Global Laboratory Alliance. This network is already operating to uniform standards, with a quality assurance system in place to maintain compliance and ensure consistency in testing globally. Although this system may not be needed long term, it serves a useful function at present.

THE FUTURE OF GMO TESTING

When one considers critically the suitability of current GMO testing methods (i.e., PCR- and immuno-based methods) one must conclude that these methods do not completely fulfill the needs of industry and regulators. The gap between need and technological capability is even greater if one considers a possible future in which hundreds or even thousands of GMOs are approved for use in human foods instead of the relative handful that are approved currently. In considering future technological developments, it is worthwhile to inventory the characteristics of what might be called "full-service" GMO testing technology to benchmark our expectations for future technologies. A full-service testing platform would require the following capabilities:

- Simultaneous analysis of hundreds of GMOs
- Highly sensitive
- Highly specific
- Accurate quantification
- Rapid
- Economical
- Flexible—effective with diverse food matrices
- Easy to use
- Portable
- Field operable
- Automatable

PCR is highly sensitive and specific, and offers reasonably good quantification in real-time format. However, it fails to meet some of the key requirements listed previously. It would be costly and cumbersome to screen hundreds of GMOs by PCR. Even if one could multiplex 10 or 15 primer sets into a single reaction, screening a hundred or more GMOs becomes prohibitive, both economically and logistically. PCR has always been a technically challenging and time-consuming procedure, and the expansion of the number of GMOs on the market will further exacerbate these features.

Immunodetection methods are less expensive than PCR on a per-test basis, and offer the advantages of speed of analysis and ease of use. In the lateral flow format, these methods can also be field-operable. However, they lack the sensitivity and capacity for accurate quantification that will be required for a full-service GMO analytical system. Furthermore, immunodetection methods are not well adapted to rapid, simultaneous screening of large numbers of GMOs. The greatest limitation related to this technology is, however, the costly and time-consuming effort required to develop the hundreds of antibodies needed to create immunotests for hundreds of GMOs. In summary, both PCR and immunodetection methods will inevitably freeze up when confronted with the demand to screen very large numbers of diverse GMOs. It is clear that new technological developments are required.

Earlier in this chapter we referred to two technologies currently on the horizon, namely, biosensors and microarrays. A third contender is near-infrared (NIR) spectroscopy. The remainder of this chapter will outline briefly how these three approaches will measure up to future needs.

NEAR-INFRARED SPECTROSCOPY

The infrared (IR) spectrum of a material is determined by its chemical composition. NIR is distributed widely and used routinely in the grain industry to assess the oil and moisture content of grains and oilseeds. NIR is also rapid, economical, easy to use, field operable, and automatable. Thus NIR would be advantagous if it could be adapted to distinguish between genetically modified and traditional varieties of soy, maize, and oilseed rape.

This question led to research that has demonstrated that somewhat more-sophisticated NIR equipment than that which is currently operating in the grain system can, with proper programming, differentiate between lots of soybeans that are predominately genetically modified (e.g., contain Roundup Ready traits) and lots that are primarily unmodified (Hurburgh et al., 2000). It is unfortunate, however, that the sensitivity and quantitative accuracy of this method is quite low. To be useful for addressing questions regarding GMO content that are relevant to the existing thresholds around the world (1 percent EU and Switzerland, 2 percent Norway, 3 percent Korea, 5 percent Japan) a method must be capable of detecting and quantifying a few percent of genetically modified material. NIR spectroscopy is unable to fulfill these requirements. Also, to date, success has not been forthcoming in developing similar methods for detecting other GMOs. The challenge is in accumulating an adequate database from which to glean a common NIR "signature" for a given GMO. Another level of complexity arises when analysis of real-world samples containing multiple GMOs is required. Although it may demonstrate utility as computer power linked to NIR analysis increases, thereby enabling more precise and sophisticated signal analysis, the capacity is not there at present.

DNA MICROARRAYS

Microarrays possess a number of characteristics that appear to suggest promise for future GMO detection. Consequently, some work has been carried out attempting to use microarrays for GMO detection. The forte of the microarray is the ability to detect thousands of genetic target sequences simultaneously with high specificity. However, current microarray technology is limited in sensitivity. As pointed out in a pioneering article on the use of microarrays for GMO detection (Kok et al., 2002), current methods are not sufficiently sensitive to allow detection of directly labeled GMO-specific nucleic acid targets. Instead, it is necessary to label and amplify the GMO target sequences simultaneously using PCR to achieve detectable signals.

The use of PCR in conjunction with the microarray brings to the microarray technology essentially all of the limitations mentioned previously for PCR. First, the ability to quantify is lost, since the endpoint PCR reaction, used in labeling target molecules in preparation for microarray analysis, is strictly qualitative. Second, PCR slows the analytical time frame. Chip hybridization usually requires hours. Adding a PCR amplification step to the procedure adds additional hours. Dependence on PCR also significantly reduces ease of use and makes it unlikely that portable or field-operable units could be developed. A serious limitation of linking microarrays to PCR amplification is that it makes it necessary to carry out a separate amplification reaction for every GMO of interest. That is, if one wishes to screen for 100 GMOs, it is necessary to carry out PCR amplification reactions with 100 primer sets. These reactions can be multiplexed to reduce the number of reactions carried out. However, many reactions will still be required, adding complexity to the system, and increasing the risk of operator error. Multiplexing can also generate artifacts and reduce predictability of the PCR system.

Can the microarray format be improved upon to achieve sufficient sensitivity to make it unnecessary to employ PCR amplification as a preanalytical step? Fundamental limitations of the microarray sensitivity arise from the small surface area of each micron-dimensioned probe site on the chip relative to the chip's large total surface area. When only very small numbers of target molecules are present in the solution to be analyzed, which is the case if the method is to operate successfully at the limits of detection currently required in the marketplace for GMO analysis (at least as low as 0.01 percent), then the time required for those targets to "find" their complementary probe molecules immobilized to a single, tiny spot on the chip becomes impractically long. Because of these limiting geometries, it is unlikely that microarrays will ever serve in GMO detection without use of a powerful amplification process, such as PCR. Although microarrays may not fulfill the long-term needs of GMO analysis, their advantages have led to development of microarrays for the qualitative screening of multiple GMOs. Two commercial microarray designs for detection of GMOs are currently being beta-tested (Genetic ID, n.d.; Grohmann, 2002).

The final limitation that must be overcome for making microarrays practical for routine GMO screening is economics. In their most prevalent form, microarrays are research tools, capable of screening simultaneously for many thousands of gene targets. They have found valuable application in gene expression profiling and other research applications. For such purposes, it is not a serious limitation that a single chip may cost in excess of $100, and that the instrument used to read the chip might cost in the range of $60,000 to $120,000. However, for routine GMO analysis, such costs are prohibitive.

One solution (Grohmann, 2002) is to employ a microarray format that exploits the advantages of this technology, yet minimizes the cost of the system. The key innovation is a signal visualization system that does not rely on fluorescence, which would necessitate the use of an expensive scanner, but uses an enzyme-linked system to generate a dye that is visible to the naked eye. Moreover, the dimensions of the spots where probes are attached to the chip are not a few microns, as is typical for research microarrays, but 1 to 2 mm in diameter. Thus positive hybridization signals can be clearly seen with the naked eye, hence the name EasyRead GMO Chip. The advantages are clear from Table 7.6, which compares the costs of analyzing PCR amplification products by electrophoresis, typical microarrays, and the EasyRead GMO Chip. Because of its low overhead and simple format, this system is expected to find considerable applicability in industry quality assurance labs, where a system that minimizes equipment costs and method complexity is of interest for in-house screening purposes.

TABLE 7.6. Comparison of PCR amplification analysis costs

	Electrophoresis	Typical microarray	EasyRead GMO Chip
Equipment costs	$15,000 to $20,000 (electrophoresis and imaging equipment)	$60,000 to $120,000 (microarray scanner)	$0.00
Expendable supplies costs	$0.25	$0.25	$0.25
Chip costs	$0.00	$40 to $150	$12

BIOSENSORS

Biosensors have not found routine use in GMO testing to date. However, three different kinds of biosensors have been evaluated for their suitability. These include surface plasmon resonance (SPR) (Feriotto et al., 2002; Mariotti et al., 2002; Minnuni et al., 2001), piezoelectric, and electrochemical (Minnuni et al., 2001) biosensors. When used in conjunction with PCR amplification, all three approaches were found to provide technically adequate levels of detection (Feriotto, 2002; Mariotti et al., 2002; Minnuni et al., 2001). Many other features of these biosensors offer significant advantages for GMO detection. They can be multiplexed to screen for many targets simultaneously. Their detection process is nucleic acid hybridization, which is highly selective. Other work has shown that they can function quantitatively, and because they operate on simple physical principles, detection is rapid and economical. Moreover, commercial instrumentation based on these biosensors should be easy to use and automatable. In some cases, with further work, portability and field operability should be achievable as well.

As with microarrays, sensitivity of detection is the primary limitation with the biosensors evaluated to date. It may not be possible to upgrade the sensitivity of the biosensors to achieve the sensitivity required for stand-alone use independent of PCR. However, we anticipate that within the next few years application of innovative designs and detection principles may yield other kinds of biosensors having sufficient sensitivity to adequately fulfill future GMO analytical requirements.

REFERENCES

Anklam, E. (1999). The validation of methods based on polymerase chain reaction for the detection of genetically modified organisms in food. *Anal. Chim. Acta* 393:177-179.

Anklam, E., Gadani, F., Heinze, P., Pijnenburg, H., and Van den Eede, G. (2002). Analytical methods for detection and determination of genetically modified organisms in agricultural crops and plant-derived food products. *Eur. Food Res. Technol.* 214:3-26.

Anklam, E., Heinze, P., Kay, S., Van den Eede, G., and Popping, B. (2002). Validation studies and proficiency testing. *J. of AOAC Int.* 85:809-815.

Bertheau, Y., Diolez, A., Kobilinsky, A., and Magin, K. (2002). Detection methods and performance criteria for genetically modified organisms. *J. of AOAC Int.* 85:801-808.

Brodmann, P., Eugster, A., Hübner, P., Meyer, R., Pauli, U., Vogeli, U., and Lüthy, J. (1997). Detection of the genetically engineered Roundup Ready soybeans using the polymerase chain reaction (PCR). *Mitteilungen aus dem Gebiete der Lebensmitteluntersuchung und Hygiene* 88:722-731.

Brodmann, P.D., Ilg, E.C., Berthoud, H., and Herrmann, A. (2002). Real-time quantitative polymerase chain reaction methods for four genetically modified maize varieties and maize DNA content in food. *J. of AOAC Int.* 85:646-653.

Csaikl, U.M., Bastian, H., Brettschneider, R., Gauch, S., Meir, A., Schauerte, M., Scholz, F., Sperisen, C., Vornam, B., and Ziegenhagen, B. (1998). Comparative analysis of different DNA extraction protocols: A fast, universal maxi-preparation of high quality plant DNA for genetic evaluation and phylogenetic studies. *Plant Molecular Biology Reporter* 16:69-86.

Comité Européen de Normalisation/International Organization for Standardization (CEN/ISO) (2002a). Methods for the detection of genetically modified organisms and derived products—Qualitative nucleic acid based methods. *pr EN ISO 21569 Foodstuffs*, ISO/DIS 21569:2002.

Comité Européen de Normalisation/International Organization for Standardization (CEN/ISO) (2002b). Methods of analysis for the detection of genetically modified organisms and derived products—Nucleic acid extraction. *pr EN ISO 21571 Foodstuffs*, ISO/DIS 21571:2002.

Comité Européen de Normalisation/International Organization for Standardization (CEN/ISO) (2002c). Nucleic acid based methods of analysis for the detection of genetically modified organisms and derived products—General requirements and definitions. *pr EN ISO 24276 Foodstuffs*, ISO/DIS 24276:2002.

Davies, M.J., Taylor, J.I., Sachsinger, N., and Bruce, I.J. (1998). Isolation of plasmid DNA using magnetite as a solid-phase adsorbent. *Analytical Biochemistry* 262:92-94.

Demay, F., Darche, S., Kourilsky, P., and Delassus, S. (1996). Quantitative PCR with colorimetric detection. *Research in Immunology* 147:403 409.

European Commission (EC) (1997). Council Regulation No 258/97 of the European Parliament and of the Council of 27 January 1997, concerning novel foods and novel food ingredients, OJ L 043, 14.2.1997, p.1-6.

European Commission (EC) (1998). Council Regulation No 1139/98 of 26 May 1998 concerning the compulsory indication of the labelling of certain foodstuffs produced from genetically modified organisms of particulars other than those provided for in Directive 79/112/EEC, OJ L 159, 3.6.1998, p. 4-7.

Feriotto, G., Borgatti, M., Mischiati, C., Bianchi, N., and Gambari, R. (2002). Biosensor technology and surface plasmon resonance for real-time detection of genetically modified Roundup Ready soybean gene sequences. *Journal of Agricultural and Food Chemistry* 50:955-962.

Genetic ID (n.d.). EasyRead GMO Chips. <http://www.genetic-id.com>.

German Federal Foodstuffs Act (1998). Detection of genetic modification of soybeans by amplification of the modified DNA sequence by using the polymerase chain reaction (PCR) and hybridization of the PCR product with a DNA probe. *German Federal Foodstuffs Act—Food Analysis Article 35, L23.01.22-1, Beuth, Berlin Koln.*

Greiner, R., Konietzny, U., and Jany, K.D. (1997). Is there any possibility of detecting the use of genetic engineering in processed foods? *Zeitscrift für Ernahrungswiss* 36:155-160.

Grohmann, L. (2002). GMO chip internal validation results. Symposium—GMO Analytik Heute, Frankfurt, Germany, January.

Hemmer, W. (1997). *Foods derived from genetically modified organisms and detection methods.* Basel, Switzerland: Agency for Biosafety Research and Assessment of Technology Impacts of the Swiss Priority Programme Biotechnology of the Swiss National Science Foundation. BATS-Report, February.

Hemmer, W. and Pauli, U. (1998). Labeling of food products derived from genetically engineered crops. *Eur. Food Law Rev.* 1:27-38.

Higuchi, R., Dollinger, G., Walsh, P.S., and Griffith, R. (1992). Simultaneous amplification and detection of specific DNA sequences. *Biotechnology (NY)* 10: 413-417.

Hoef, A.M., Kok, E.J., Bouw, E., Kuiper, H.A., and Keijer, J. (1998). Development and application of a selective detection method for genetically modified soy and soy-derived products. *Food Additives and Contaminants* 15:767-774.

Hübner, P., Studer, E., Hafliger, D., Stadler, M., Wolf, C., and Looser, M. (1999a). Detection of genetically modified organisms in food: Critical points for quality assurance. *Accreditation and Quality Assurance* 4:292-298.

Hübner, P., Studer, E., and Lüthy, J. (1999a). Quantitation of genetically modified organisms in food. *Nature Biotechnology* 17:1137-1138.

Hübner, P., Studer, E., and Lüthy, J. (1999b). Quantitative competitive PCR for the detection of genetically modified organisms in food. *Food Control* 10:353-358.

Hurburgh, C.R., Rippke, G.R., Heithoff, C., Roussel, S.A., and Hardy, C.L. (2000). Detection of genetically modified grains by near-infrared spectroscopy. *Pro-

ceedings PITTCON 2000—Science for the 21st Century, #1431, New Orleans, LA. March 12-17.

International Organization for Standardization (ISO) (1999). ISO Guide 17025.

Jaccaud, E., Hohne, M., and Meyer, R. (2003). Assessment of screening methods for the identification of genetically modified potatoes in raw materials and finished products. *Journal of Agricultural and Food Chemistry* 51:550-557.

Jankiewicz, A., Broll, H., and Zagon, J. (1999). The official method for the detection of genetically modified soybeans (German Food Act LMBG 35): A semi-quantitative study of sensitivity limits with glyphosate-tolerant soybeans (Roundup Ready) and insect-resistant Bt maize (Maximizer). *European Food Research and Technology (Zeitschrift fur Lebensmittel-Untersuchung und-Forschung A)* 209:77-82.

Japanese Ministry of Health Labour and Welfare (JMHLW) (2002a). Genetically Modified Foods <http://www.mhlw.go.jp/english/topics/qa/gm-food/ index.html>.

Japanese Ministry of Health Labour and Welfare (JMHLW) (2002b). Testing for Foods Produced by Recombinant DNA Techniques <http://www.mhlw.go.jp/ english/topics/food/sec05-1a.html>.

Joint Research Centre of the European Commission (JRC-EC) (2002). European Network of GMO Laboratories (ENGL). <http://engl.jrc.it>.

Kay, S. and Van den Eede, G. (2001). The limits of GMO detection. *Nature Biotechnology* 19:405.

Kok, E.J., Aarts, H.J.M., van Hoef, A.M.A., and Kuiper, H.A. (2002). DNA methods: Critical review of innovative approaches. *J. of AOAC Int.* 85:797-800.

Lipp, M., Anklam, E., and Stave, J.W. (2000). Validation of an immunoassay for detection and quantitation of a genetically modified soybean in food and food fractions using reference materials: Interlaboratory study. *J. of AOAC Int.* 83: 919-927.

Lipp, M., Brodmann, P., Pietsch, K., Pauwels, J., and Anklam, E. (1999). IUPAC collaborative trial study of a method to detect genetically modified soy beans and maize in dried powder. *J. AOAC Int.* 82:923-928.

Lipton, C.R., Dautlick, J.X., Grothaus, G.D., Hunst, P.L., Magin, K.M., Mihaliak, C.A., Rubio, F.M., and Stave, J.W. (2000). Guidelines for the validation and use of immunoassays for determination of introduced proteins in biotechnology enhanced crops and derived food ingredients. *Food and Agricultural Immunology* 12:153-164.

Lüthy, J. (1999) Detection strategies for food authenticity and genetically modified foods. *Food Control* 10:359-361.

Mariotti, E., Minunni, M., and Mascini, M. (2002). Surface plasmon resonance biosensor for genetically modified organisms detection. *Analytica Chimica Acta* 453:165-172.

Melzak, K., Sherwood, C., Turner, R., and Haynes, C. (1996) Driving forces for DNA adsorption to silica in perchlorate solutions. *Journal of Colloid and Interface Science* 181:635-644.

Meyer, R. (1999). Development and application of DNA analytical methods for the detection of GMOs in food. *Food Control* 10:391-399.

Minunni, M., Tombelli, S., Mariotti, E., and Mascini, M. (2001). Biosensors as new analytical tool for detection of genetically modified organisms (GMOs). *Fresenius Journal of Analytical Chemistry* 369:589-593.

Murray, M.G. and Thompson, W.F. (1980). Rapid isolation of high molecular weight plant DNA. *Nucleic Acids Research* 8:4321-4325.

New Zealand Ministry of Agriculture and Forestry (NZMAF) (2002). Protocol for testing imports of *Zea mays* seed for sowing for the presence of genetically modified seed. <http://www.maf.govt.nz/biosecurity/imports/plants/seeds-sowing.htm>, 1-5.

Pietsch, K., Bluth, A., Wurz, A., and Waiblinger, H.U. (1999). PCR method for the quantification of genetically modified soybeans. *Deutsche Lebensmittel-Rundschau* 95:57-59.

Pietsch, K., Waiblinger, H.U., Brodmann, P., and Wurz, A. (1997). Screening methods for identification of "genetically modified" food of plant origin. *Deutsche Lebensmittel-Rundschau* 93:35-38.

Schweizerisches Lebensmittelbuch (Chapter 52B) (1998). Molekularbiologische Methoden in Bundesamt fur Gesundheit Eidgenossische Drucksachen und Materialzentrale, Bern, Switzerland.

Scott, O.R. and Benedich, A.J. (1988). Extraction of DNA from plant tissues. *Plant Molecular Biology Manual* A6, 1-10.

Shena, M. (ed.) (2000) *Microarray Biochip Technology*. Eaton Publishing, Nattick, MA.

Shindo, Y., Kuribara, H., Matsuoka, T., Futo, S., Sawada, C., Shono, J., Akiyama, H., Goda, Y., Toyoda, M., and Hino, A. (2002). Validation of real-time PCR analyses for line-specific quantitation of genetically modified maize and soybean using new reference molecules. *J. AOAC Int.* 85:1119-1126.

Shirai, N., Momma, K., Ozawa, S., Hashimoto, W., Kito, M., Utsumi, S., and Murata, K. (1998). Safety assessment of genetically engineered food: Detection and monitoring of glyphosate-tolerant soybeans. *Bioscience, Biotechnology, and Biochemistry* 62:1461-1464.

Stave, J.W. (1999). Detection of new or modified proteins in novel foods derived from GMO—Future needs. *Food Control* 10:367-374.

Stave, J.W. (2002). Protein immunoassay methods for detection of biotech crops: Applications, limitations, and practical considerations. *J. AOAC Int.* 85:780-786.

Taoufik, Y., Froger, D., Benoliel, S., Wallon, C., Dussaix, E., Delfraissy, J.F., and Lantz, O. (1998). Quantitative ELISA-polymerase chain reaction at saturation using homologous internal DNA standards and chemiluminescence revelation. *Eur. Cytokine Network* 9:197-204.

Terry, C.F., Harris, N., and Parkes, H.C. (2002). Detection of genetically modified crops and their derivatives: Critical steps in sample preparation and extraction. *J. AOAC Int.* 85:769-774.

Tozzini, A.C., Martinez, M.C., Lucca, M.F., Rovere, C.V., Distefano, A.J., del Vas, M., and Hopp, H.E. (2000). Semi-quantitative detection of genetically modified grains based on CaMV 35S promoter amplification. *Electronic J. of Biotechnol.* 3:1-6.

Trapmann, S., Schimmel, H., Kramer, G.N., Van den Eede, G., and Pauwels, J. (2002). Production of certified reference materials for the detection of genetically modified organisms. *J. AOAC Int.* 85:775-779.

Tyagi, S. and Kramer, F.R. (1996). Molecular beacons: Probes that fluoresce upon hybridization. *Nature Biotechnology* 14:303-308.

Vaitilingom, M., Pijnenburg, H., Gendre, F., and Brignon, P. (1999). Real-time quantitative PCR detection of genetically modified maximizer maize and Roundup Ready soybean in some representative foods. *Journal of Agricultural and Food Chemistry* 47:5261-5266.

Van den Eede, G., Kay, S., Anklam, E., and Schimmel, H. (2002). Analytical challenges: Bridging the gap from regulation to enforcement. *J. AOAC Int.* 85:757-761.

Vogelstein, B. and Gillespie, D. (1979). Preparative and analytical purification of DNA from agarose. *Proceedings of the National Academy of Science USA* 76: 615-619.

Vollenhofer, S., Burg, K., Schmidt, J., and Kroath, H. (1999). Genetically modified organisms in food-screening and specific detection by polymerase chain reaction. *Journal of Agricultural and Food Chemistry* 47:5038-5043.

Wassenegger, M. (1998). Application of PCR to transgenic plants. *Methods Molecular Biology* 92:153-164.

Whitaker, T.B., Springer, J., Defize, P.R., deKoe, W.J., and Coker, R. (1995). Evaluation of sampling plans used in the United States, United Kingdom, and the Netherlands to test raw shelled peanuts for aflatoxin. *J. AOAC Int.* 78:1010-1018.

Whitecomb, D., Theaker, J., Guy, S.P., Brown, T., and Little, S. (1999). Detection of PCR products using self-probing amplicons and fluorescence. *Nature Biotechnology* 17:804-807.

Wiseman, G. (2002). State of the art and limitations of quantitative polymerase chain reaction. *J. AOAC Int.* 85:792-796.

Wittwer, C.T., Herrmann, M.G., Moss, A.A., and Rasmussen, R.P. (1997). Continuous fluorescence monitoring of rapid cycle DNA amplification. *BioTechniques* 22:130-131, 134-138.

Wurz, A., Bluth, A., Zeltz, P., Pfeifer, C., and Willmund, R. (1999). Quantitative analysis of genetically modified organisms (GMO) in processed food by PCR-based methods. *Food Control* 10:385-389.

Zimmermann, A., Lüthy, J., and Pauli, U. (1998). Quantitative and qualitative evaluation of nine different extraction methods for nucleic acids on soybean food samples. *Z. Lebensum. Unters. Forsch. A.* 207:81-90.

Chapter 8

DNA-Based Methods
for GMO Detection: Historical
Developments and Future Prospects

Farid E. Ahmed

Chapter 7 discussed the principles of DNA-based methods for GMO detection and related quality control/assurance issues. This chapter shall extend the discussion, outlining major historical developments in testing, giving specific examples of PCR tests used as official methods in some countries to meet labeling regulations, and elaborating further on other DNA-based techniques that may be of research interest, such as Southern blotting, a high throughput method—DNA microarray technology—that may have application to GMO testing, and other real-time biosensor assays that are applicable to field-testing of GMOs.

SOUTHERN HYBRIDIZATION/SLOT
OR DOT BLOTTING TECHNIQUES

Southern blotting is a localization technique for a particular sequence within the DNA (Southern, 1975), in which the DNA fragments are separated according to size by gel electrophoresis, followed by in situ denaturation and transfer from the gel (by capillary, vacuum, or electrophoresis) to a solid support (e.g., nitrocellulose or nylon membrane). If nitrocellulose filters are used, they are usually

---

I wish to thank those colleagues who kindly provided me with the figures and insight about the application of new technologies, considerations that provided an in-depth analysis enhancing the readability and clarity of this chapter.

baked at 80°C for 2 h. However, if durable nylon membranes are used, they must be treated to fix (immobilize) DNA either by drying under vacuum or by exposure to low doses of 254 nm of ultraviolet light (UV) to attach DNA to the membrane covalently. DNA is then hybridized to a probe (from 17 to several hundred nucleotides long) that is radiolabeled or attached to other substances for colorimetric or chemiluminescence detection using techniques such as nick translation or end labeling (Hill et al., 1999; Rivin et al., 1986). Formamide is usually added to hybridization solution for radiolabeled probes, and background binding of the probe may be suppressed by including in the hybridization medium either 1 percent sodium dodecyl sulfate (SDS), Denhardt's reagent or 0.05 × BLOTTO (Denhardt, 1966; Johnson et al., 1984). Southern hybridization is so sensitive when ^{32}P is the radiolabel (specific activity > 10^9 cpm/µg) that a sequence of 1,000 bp can be detected in an overnight exposure of ~10 µg DNA (~30 µg DNA is needed if the probe is <20 nucleotides long) (Ikuta et al., 1987). The strength of the signal is proportional to the specific activity of the probe and inversely proportional to its length, reaching the limits of its intensity when very short probes are used (Sambrook and Russel, 2000).

Dot blots or slot blots are used to facilitate loading several heat-denatured DNA samples (95°C for 12 min and 0.2 M NaOH) onto slots of a microfiltration acrylic manifold (Figure 8.1a) using a positively charged nylon membrane, after which DNA is cross-linked by exposure to UV, in the case of a nylon membrane, or by baking for 2 h at 80°C, in the case of a nitrocellulose filter, followed by hybridization with a digoxygenin-labeled probe and viewing (Figure 8.1b) (Rivin et al., 1986). Use of nonradioactive probes (e.g., digoxygenin or biotin) has gained popularity as recent labeling developments achieved sensitivity that nearly matches radioactive method (Ross et al., 1999).

Nonradioactive probe labeling methods are separated into two different approaches: indirect and direct (Table 8.1). Indirect, hapten-based labeling kits, such as Gene Images Labeling and Detection Systems (Amersham Biosciences), introduce nucleotides tagged with fluorescein into the probe. These are then detected with a highly specific antifluorescein antibody conjugated to the enzyme alkaline phosphatase (AP) (Figure 8.2a). Fluorescein-labeled DNA is stable under standard hybridization conditions, as with radiolabeled probes, and

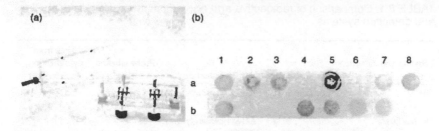

FIGURE 8.1. Dot/slot blotting apparatus and its use in the analysis of *Penicillium nalgiovense* transformants. (a) Dot blot microfiltration manifold. (b) Dot blot analysis of *P. nalgiovense*. The DNA of 14 ATCC66742 cotransformants, transformed with equal amounts of p3SR2 and pELN5-lac, was isolated. The DNA was resuspended in 100 μl TE, heated at 95°C for 10 min and cooled immediately on ice. The DNA was tranferred to nitrocellulose filters, which were baked for 2 h at 80°C and prehybridized for 2 h at 68°C. A 3.0 kb DNA fragment carrying part of the *Escherichia coli lacZ* gene was digoxygenin labeled and used as a hybridization probe. Hybridization was carried out at 68°C overnight, followed by washing and developing. The dots (1-7, a + b) represent the analyzed samples. As a positvie control (8a), 10 ng of pELN5-lac, and as a negative control (8b), the DNA of the untransformed strain was used. (*Source:* Reprinted from Rivin et al., 1986, with permission.)

the stringency of hybridization can be controlled with either temperature or salt concentration. Chemiluminescent detection using CDP-Star (Amersham Biosciences) as the substrate for the enzyme enables as little as 50 fg of target DNA to be detected, making it a sensitive alternate to ^{32}P (Table 8.1), is capable of detecting 10 fg of target DNA using nucleotides labeled with a ^{32}P highly specific probe up to 2×10^9 dpm/μg. Blots hybridized using Gene Images Random-Prime Labeling and 3'-oligolabeling modules can also be detected with ECF substrates (Amersham Biosciences). The nonfluorescent substrate is catalyzed by AP to produce a chemifluorescent signal that accumulates over time. Low sensitivity applications yield results after 1 h, whereas high sensitivity applications usually require overnight incubation. Quantification of low-sensitivity applications such as dot and slot blots is possible through ECF detection, and the results can be analyzed using various imaging systems that allow filmless detection and subsequent image analysis employing powerful software packages, such as Image Manager and Image Quant (Ross et al., 1999).

TABLE 8.1. Comparison of radioactive and nonradioactive nucleic acid labeling and detection systems

Labeling and detection system	Sensitivity	Application	Probe labeling time	Time from hybridization to detection
32p	Down to 10 fg	All high-sensitivity applications	5 min to 3 h	On film 1 h to 1 week
AlkPhos Direct	60 fg	All high-sensitivity applications	30 min	1 h
ECL Direct	0.5 pg	High-target applications	20 min	1-2 h
Gene Images Random-Prime with CDP-Star Detection	50 fg	High-sensitivity Northern blots	30 min	3 h
Gene Images 3'-Oligolabeling with CDP-Star	0.1 pg	Oligonucleotide screening with stringency control	30 min	3 h
Gene Images Random-Prime with ECF	0.25 pg	Quantification	30 min	3 h
Gene Images 3'-Oligolabeling with ECF	120 pg	Quantification	30 min	3 h
ECL Random-Prime	0.5 pg	Medium-target Southern blots with DNA probes	30 min	3 h
ECL 3'-End Labeling	0.2 pg	Medium- to high-target Southern blots with oligonucleotide probes	30 min	3 h

Source: Reproduced from Osborn, 2000, with kind permission of Amersham Biosciences Corp.

Direct labeling methods are alternatives to indirect methods, and offer significant improvements in speed and convenience without compromising sensitivity. CyDye (Molecular Probes) labels are fluorescent dyes that can be coupled directly to nucleotides or conjugated to antibodies. Fluorescently labeled nucleotides are also available, and these can be incorporated into DNA probes by nick translation or random priming. The fluorophores of hybridized probes can then be visualized directly using fluorescent scanners. With direct chemifluorescence labeling reagents such as AlkPhos Direct (Amersham Biosciences), AP is directly crosslinked to the nucleic acid probe in a simple, 30-min reaction (Figure 8.2b). The probe is hybridized to the blot and incubated. Detection is possible 1 h after hybridization. Di-

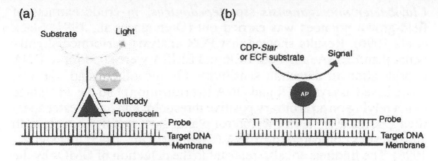

FIGURE 8.2. (a) Outline for the indirect probe labeling method and (b) the direct labeling method (*Source:* Reproduced from Osborn, 2000, with kind permission of Amersham Biosciences Corp.)

rect labeling saves 3 to 4 h compared to indirect labeling, because the antibody-conjugate incubation and associated blocking and washing steps are eliminated (Osborn, 2000).

Chemiluminescent detection systems were originally used with autoradiographic films because of the low levels of light produced. However, imaging platforms incorporating highly sensitive optic systems, such as Typhoon 8600 and Imager Master DS-CL, enable the direct detection of chemiluminescence without the need for intermediate exposure to autoradiographic films or screens. However, unlike the signal produced by a radioactive probe, chemiluminescence is linear only over a narrow range, and therefore offers limited quantification. Thus, although fluorescence, chemifluorescence, and chemiluminescence have emerged as alternative technologies to the traditional radioisotope-based systems still used in research application, because of convenience, speed and ease of automation, situations exist in which radioisotopes may have a sensitivity advantage over nonradioactive methods, and are thus chosen in spite of the hazard and inconvenience due to their use (Osborn, 2000).

A fluorescein-N^6 dATP-end-labeled probe and signal detected via antifluorescein horseradish peroxidase (HRP), when used to identify the conjugate *APH IV* gene for protoplast-derived maize by DNA hybridization, was reported to produce signals equivalent to those obtained by ^{32}P labeling (Hill et al., 1999).

A comparison of the performance of PCR, ELISA, and DNA hybridization for the detection of the causal agent of bacterial ring rot,

Clavibacter michiganensis ssp. *sepedonicus,* in crude extracts of field-grown potatoes was carried out (Drennan et al., 1993; Slack et al., 1996). Results showed that PCR analysis performed slightly better than ELISA, and both PCR and ELISA were superior to DNA hybridization in detection sensitivity. On the other hand, the two DNA-based assays (PCR and DNA hybridization) had the advantage of not relying on an arbitrary positive threshold, and had greater specificity, because none of the control plants gave positive results with either test as compared to ELISA (Drennan et al., 1993; Slack et al., 1996). The findings are also relevant to the detection of GMOs by the three methods mentioned previously.

In 2001, an alternative Southern blot technology was created using near-infrared fluorescence dyes (emissions at ~ 700 and 800 nm) that were coupled to a carbodiimide reactive group and attached directly to DNA in a 5-min reaction. The signals for both dyes were detected simultaneously (limit in the low zeptomolar range) by two detectors of an IR imager, such as the Odyssey I infrared (IR) imaging system manufactured by LI-COR Biosciences; something not yet possible with conventional radioactive or chemiluminescent detection techniques (Stull, 2001).

Although Southern hybridization can be quantitative, it is used primarily as a research tool for GMO detection. It is unsuitable for routine or automated testing because of its low throughput and high cost (two days for completion at a cost of $150/sample) (Ahmed 2002a).

POLYMERASE CHAIN REACTION APPLICATIONS

Because not all GM foods contain expressed protein(s) or have antibodies available to detect them, and because of the rather low expression levels of transgenic products in tissue used for human consumption (most are in the lower ppm or even the ppb or ppt range) (Hemmer, 1997), more sensitive PCR methods for the detection of common DNA elements, such as markers or widely used promoters and terminators, are used today to screen for several different GMO foods by more or less the same assay. A positive result, however, has to be confirmed by a specific assay determining the unique modification (Schreiber, 1999). Two essential prerequisites for the application of PCR-based detection methods are complete knowledge of the foreign gene construct within the GMO to be detected, and the ability to extract significant amounts of amplifiable DNA from the samples to be assayed either by qualitative or quantitative PCR methods (Ahmed, 1995, 2000). The limiting factors for PCR detection are the availability of certified reference material (CRM) and criteria for standardization (Wurz et al., 1999).

The first method for GMO identification in foodstuffs, developed to identify Flavr Savr tomatoes, was a PCR application assay because the genetic modification did not produce a protein in the plant (Meyer 1995). In addition to the polygalacturonase (PG) gene, which degrades pectin in the cell wall, Flavr Savr tomatoes contain the Kan^r gene, which confers resistance to kanamycin and the cauliflower mosaic virus promoter CaMV 35S. PCR detection was achieved by designing two pairs of primers: one pair amplified a 173 bp fragment for kan^r, and the second pair amplified a 427 bp fragment that contains part of the promoter sequence (Lüthy, 1999).

A set of specific and sensitive PCR primers were developed for Roundup Ready (RR) soybean containing a new genetic element from *Agrobacterium tumefaciens* producing the enzyme 5-enolpyruvylshikimate-3-phosphate synthase (EPSPS) that makes the plant resistant to the herbicide glyphosate, and the genetically modified Maximizer maize (MM) containing the synthetic endotoxin *CryIAb* gene (Gasser and Fraley, 1989). For RR, a nested PCR method setting the primers into three of four newly introduced genetic elements, namely the 5' end of the CP4 EPSPS gene, the chloroplast transit peptide gene, and the 35S promoter. The detection of this combination of genetic elements is highly specific for the RR soybean, giving a detection limit of 20 pg DNA, corresponding to 0.01 to 0.1 percent GM soybean in unmodified product (Köppel et al., 1997). For MM, primers were directed against the *CryIAb* gene, giving a detection limit comparable to RR. To control the amplification potential of the extracted DNA, a second maize-specific PCR system based on the zein storage gene was developed to check whether a maize-derived product contained modified DNA (Lüthy, 1999). An alternative to the specific PCR methods mentioned previously is a screening method that uses target sequences in genetic elements found most commonly in transgenic crops, such as the CMV 35S-promoter, the nopaline synthase (NOS) terminator, and the aminoglycoside 3'-phosphotransferase (npt II) marker gene (Pietsch et al., 1997; Vollenhofer et al., 1999).

A nested PCR method was later applied to the detection of the EPSPS gene in soy bean meal pellets and flour, as well as a number of processed complex products, such as infant formulas, tofu, tempeh, soy-based desserts, bakery products, and meal-replacing products (Van Hoef et al., 1998). In this two-step method, an outer primer was used to amplify a 352 bp fragment, followed by an inner primer set to amplify another 156 bp fragment. This resulted in improved selectivity and sensitivity of the PCR reaction. RR bean DNA could be detected at 0.02 percent, but processed products (e.g., candy, biscuits, lecithins, cocoa-drink powder, and vegetarian paste) were undetectable by PCR due to DNA breakdown as a result of heating (Figure 8.3) and low pH, which resulted in increased nuclease activity leading to depurination and hydrolysis (Gasch et al., 1997; Hupfer et al., 1998; Meyer, 1999; Wurz et al., 1999). The presence of inhibitory components and low amounts of DNA in some material (e.g., lecithin, starch derivatives and refined soybean oil) makes it difficult to develop a single reliable method for the detection of all products (Van Hoef et al., 1998). Some extraction kits, for example, QIAamp DNA Stool Mini Kit (QIAGEN Inc.), were reported to effectively remove PCR inhibitory substances, such as cocoa, from highly processed foods containing GMOs (Tengel et al., 2001).

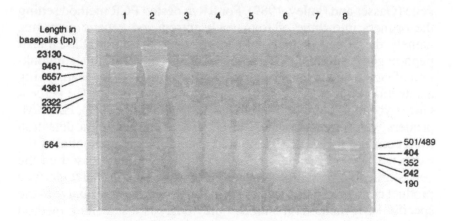

FIGURE 8.3. Shortening of DNA fragments following heat denaturation. Bovine DNA (2 µg) was employed either fresh, i.e., no heating (lane 2); treated at 100°C for 5 min (lane 3); at 100°C for 15 min (lane 4); at 100°C for 30 min (lane 5); at 121°C for 5 min (lane 6); and at 121°C for 13 min (lane 7). Size markers were used in lanes 1 and 8. (*Source:* Courtesy of GeneScan Europe, 2000.)

A semiquantitative method for RR detection based on the limited dilution method was the official method for detection of GM foods in Germany (Jankiewicz et al., 1999). This method is based on (1) optimization of the PCR so that amplification of an endogenous control gene will take place in an all-or-none fashion occurring at the terminal plateau phase of the PCR, and (2) the premise that one or more targets in the reaction mixture (e.g., GMOs) will give rise to a positive result. Accurate quantitation is achieved by performing multiple replicates at serial dilutions of the material(s) to be assayed. At the limit of dilution, where some end points are positive and some are negative, the number of targets present can be calculated from the proportion of negative end points by using Poisson statistics (Sykes et al., 1998). In this method, two measurements are used for setting limits for the GMO content of foods: a theoretical detection limit ($L_{Theoret}$) defined as the lowest detectable amplification determined from the serial dilution of target DNA with/without background DNA, and a practical detection limit (L_{Pract}) defined as the lowest detectable amplicon determined by examining certified reference material (EC, 2001) containing a different mass fraction of GM and non-GM organisms. The $L_{Theoret}$ for both RR and MM (0.0005 percent) is generally two or more orders of magnitude lower than L_{Pract} (0.1 percent) (Jankiewicz et al., 1999). An advantage of this method is that it does not require coamplification of added reporter DNA. However, caution should be exercised when using this technique because of potential contamination of PCR reactions due to various dilutions and manipulations (Hupfer et al., 1998).

As stated earlier, quantitation is an important aspect in GMO food analysis because maximum limits of GMO in food are the basis for labeling in EU countries, and the increasing number of GM foods on the market demands the development of more advanced multidetection systems (Schreiber, 1999). Therefore, adequate quantitative PCR detection methods were developed. Quantitative competitive (QC) PCR was first applied for the determination of the 35S-promoter in RR and MM in Switzerland as described by Studer et al. (1998), and is currently the official quantitative detection method for GMO-containing food in Switzerland (Swiss Food Manual, 1998). Although Switzerland is not a member of the EU, in 1999 it revised its food regulation, introducing a threshold level of 1 percent GMO content as the basis for food labeling (Hübner et al., 1999a). Many of these regulations were instituted to harmonize worldwide trade and prevent stoppage of foods, trade barriers, and retaliations among nations (Ahmed, 1999). In this quantitative endpoint PCR method (Figure 8.4), an internal DNA standard was coamplified with target DNA (Hübner et al., 1999b). The standard was constructed using linearized plasmids containing a modified PCR amplicon that was an internal insert of 21 base pair (bp) deletion or a 22 bp point mutation in the case of RR and MM DNA, respectively. These standards were calibrated by coamplification with a mixture containing different amount of RR and MM, respectively. In these systems, the presence of PCR inhibitors will be noticed immediately, since both the amplification of internal standard and target DNA will be affected simultaneously, resulting in a more robust quality control of extracted DNA, and making the system superior to noncompetitive systems and less subject to wide, interlaboratory variations (Hardegger et al., 1999; Hübner et al., 1999a,b; Meyer, 1999; Studer et al., 1998). QC-PCR consists of four steps:

1. Coamplification of standard and target DNA in the same reaction tube (Figure 8.4a)
2. Separation of the products by an appropriate method, such as agarose gel electrophoresis, and staining the gel by ethidium bromide (Figure 8.4b)
3. Analysis of the gel densitometrically (Figure 8.4c)
4. Determining the relative amounts of target and standard DNA by regression analysis (Figure 8.4d)

At the equivalent point (Figure 8.4b, iv) the starting concentration of internal standard and target are equal (i.e., the regression coefficient r^2 is greater than 0.99 and the slope of the regression line is close to unity (Raeymackers, 1993). In QC-PCR, the competition between the amplification of internal standard DNA and target DNA generally leads to loss of detection sensitivity. Nevertheless, the Swiss example depicted herein allows for the detection of as little as 0.1 percent GMO DNA (Studer et al., 1998; Hübner et al. 1999b), thus permitting the analyst to survey threshold limits in foods as specified by the European Novel Food Regulations (European Commission, 1997).

To overcome some of the limitations of conventional quantitative endpoint PCR, a real-time PCR was introduced that provided a large,

(a) Coamplification of standard and target DNA

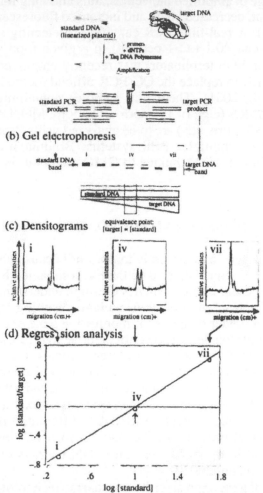

(b) Gel electrophoresis

(c) Densitograms

(d) Regression analysis

FIGURE 8.4. Schematic of milestones carried out in a Swiss QC-PCR. Standard and target DNAs are coamplified in the same reaction tube (a). Following the PCR, the products are separated by gel electrophoresis (b), which distinguishes the standard DNA from the amplified target by the size of the product. Gels were stained with ethidium bromide. At the equivalent point (iv), the starting concentrations of internal standard and the target are equivalent. Densitometric analysis of the various bands (c) can be used to calculate the linear regression (d). (*Source:* Reprinted from *Food Control*, Vol 10, Hübner et al., Quantitative competition PCR for the detection of genetically modified organisms in food, pp. 353-358, Copyright 1999, with permission from Elsevier Science.)

dynamic range of amplified molecules, thus allowing for higher sample throughput, decreased labor, and increased fluorescence (Ahmed, 2000, 2002b). A real-time PCR capable of detecting several or all GMOs (between <0.1 to >1 percent) in a given food sample (e.g., CaMV 35S or NOS terminator) was recently employed in Switzerland and the EU to replace the QC-PCR official method of detection (Hübner et al., 2001). Thereafter, samples containing substantial amounts of GMOs (e.g., Bt 176 corn, Bt 11 corn, MON 810 corn, GA 21 corn, BH 351 corn, etc.) are to be analyzed using "specific" quantitative real-time multiplex PCR systems (Brodmann et al., 2002). The application of this technique to GMO quantitation was detailed in Chapter 7.

GENE EXPRESSION AND MICROARRAY TECHNOLOGY

Analysis of gene expression is important because changes in the physiology of an organism or a cell are accompanied by changes in patterns of gene expression, which is useful in understanding the consequences of genetic modification in plants resulting from the presence of GMOs. Older methods for monitoring gene expression, such as Northern blots, nuclease protection, plaque hybridization, and slot blots, are inherently serial, require large samples, measure a single messenger ribonucleic acid (mRNA) at a time, and are difficult to automate. Some more recent techniques for the analysis of gene expression include (1) comprehensive open systems as exemplified by serial analysis of gene expression (SAGE), differential display (DD) analysis, and real-time (RT)-PCR, and (2) focused closed systems, such as several forms of high-density cDNA arrays or oligonucleotide chips (Ahmed, 2002b).

The study of gene expression by microarray technology—which is still developing—is based on immobilization of cDNA or oligonucleotides on solid support. In the cDNA approach (Figure 8.5a), cDNA (or genomic clones) are arranged in a multiwell format and amplified by PCR, often using biotinylated primers. Products of this amplification—about 500-2000 base pair (bp) clones from the 3' region of studied gene—are spotted onto a solid support using high-speed robotics (Schummer et al., 1997). New scanners and spotters that increase the array's surface area to volume decrease hybridization times and increase signal intensity (Constans, 2003c).

(a) cDNA Microassays

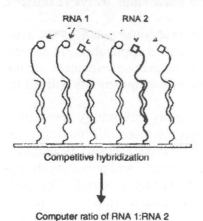

(b) Oligonucleotide GeneChips

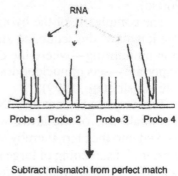

FIGURE 8.5. A schematic comparison of a cDNA and an oligonucleotide microarray. (a) In cDNA microarray analysis, nanoliter amounts of concentrated 1 to 2 kb PCR reaction product deposited on a glass or filter. These products are hybridized competitively to fluorescently labeled cDNA derived from two different RNA sources, and the ratio of the two signals at each spot reflects the relative levels of transcript abundance. (b) GeneChip consists of ~20 microsquares of 25 mer oligonucleotides per gene, including perfect and mismatch pairs (not shown) that will hybridize specifically or nonspecifically to a different part of the same transcript. Each square yields a different intensity measure, reflecting differences in GC content and folding of the RNA, so the values are massaged to produce a measure of gene expression that is contrasted with measures from other chips. (*Source:* Reprinted from *Molecular Ecology,* Vol. 11, Gibson, G., Microarrays in ecology and evolution: A preview, pp. 17-24, Copyright 2002, with permission from Blackwell Publishing Ltd.)

Oligonucleotide arrays (Figure 8.5b) are constructed by spotting prefabricated in situ oligos on a glass surface photolithographically (Pease et al., 1994) or by utilizing a piezoelectric printing method (Nuwaysir et al., 1999). Oligos ranging from 30 to 50 bases long are produced. The utility of this approach lies in its ability to discriminate between DNA molecules based on a single base-pair difference (Ahmed 2002b). Amersham Biosciences' CodeLink, a new type of array that eliminates hybridization problems, puts the oligo probe into a three-dimensional gel-like matrix that mimics a solution-phase environment and represents each gene with a functionally validated

30-mer sequences. This array offers higher reproducibility, specificity, and sensitivity than the standard glass slide arrays (Constans, 2003c).

The complexity of the hybridization reaction depends on the average length of the sequence and the intended application. In expression monitoring—where the details of the precise sequences are unimportant—sets of nucleotides that identify unique motifs will suffice (Nuwaysir et al., 1999)

The major advances of DNA microarray technology result from the small size of the array which permits more information to be packed into the chip, thereby allowing for higher sensitivity, enabling the parallel screening of large number of genes, and providing the opportunity to use smaller amounts of starting material (Constans, 2003c). The introduction of differently labeled fluorescent probes (e.g., Cy3-dUTP and Cy5-dUTP, Molecular Probes) for control and test samples has made miniaturization of arrays possible (Jordan, 1998). In addition, the production of microarrays in series facilitates comparative analysis of a number of samples (Cortese, 2000). Different matrixes have been employed for array manufacturing:

1. Glass (for short and long nucleotides and PCR products)
2. Membrane (for PCR products)
3. Microelectronics (for 10 to 400 bp fragments)
4. Polyacrylamide for oligonucleotide microchips (Ye et al., 2001)

Figure 8.5 illustrates differences between cDNA and oligonucleotide microarray technologies (Gibson, 2002).

Fluorescent signals are detected using a custom-designed scanning confocal microscope equipped with a motorized stage and laser for flour excitation (Schena et al., 1995). The generated data are analyzed with custom digital image analysis software that determines the ratio of flour 1 to flour 2 for each DNA (Ermolaeva et al., 1998). Figure 8.6 illustrates the principal of microarray amplification and analysis, whereas Figure 8.7 depicts results of a microarray experiment in the bacterium *Bacillus subtilis* (Ye et al., 2001). Powerful mining of microarray data is being developed for the assessment of statistical significance of gene expression differences, as well as the parsing of variance components due to different factors under consideration. Quantitative genomics using microarray or oligonucleotide array an-

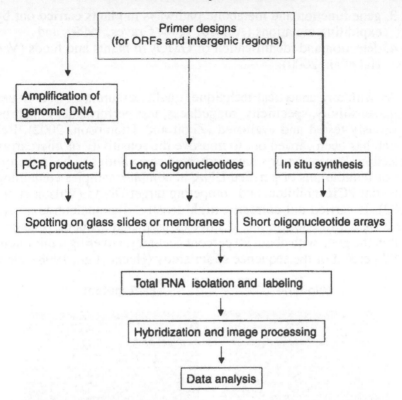

FIGURE 8.6. Construction of a DNA microarray showing principles for gene expression analysis (*Source:* Reprinted from *Journal of Microbiological Methods*, Vol. 47, Ye et al., Application of DNA micorarrays in microbial systems, pp. 257-272, copyright 2001, with permission from Elsevier.)

alysis offers the ability to estimate fundamental parameters of gene expression variation, including the additivity, dominance, and heritability of transcription (Gibson, 2002).

In the food safety arena, microarray technology has been used for applications such as

1. safety assessment of genetic modification of food (Doblhoff-Dier et al., 1999),
2. functionality of food components (Ruan et al., 1998; Kuipers, 1999),

3. gene function and metabolic pathways in plants carried out by exploiting mutations (Ahmed, 1998; Graves, 1999), and
4. detection and identification of GMOs in plants and foods (Van Hal et al., 2000).

As with any analytical technique, quality assurance (QA) issues (e.g., sensitivity, specificity, ruggedness, test performance) must be rigorously tested and evaluated (Zhou and Thompson, 2002). Research has been carried out to improve the sensitivity of microarray detection. Recent reports have indicated sensitivities to be in the order of femtograms of purified DNA in complex samples containing potential PCR inhibitors and competing target DNAs (Wilson et al., 2002). If carried out correctly, high-density oligonucleotide arrays were shown to confirm the identity of predicted sequence changes within the gene with about 99 percent accuracy, covering a minimum of 97 percent of the sequence under study (Hacia et al., 1998). New

Wild type   **FNR mutant**

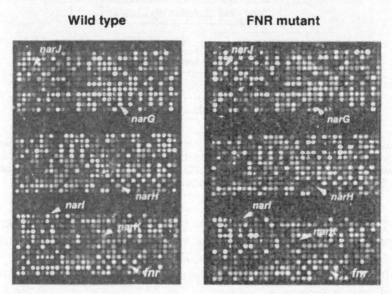

FIGURE 8.7. Illustration of a DNA microarray experiment in the bacterium *Bacillus subtilis*. This image shows the induction of *nar* genes involved in nitrate reduction and their regulation by FNR protein under anaerobic conditions. (*Source:* Reprinted from *Journal of Microbiological Methods,* Vol. 47, Ye et al., Applications of DNA microarrays in microbial systems, pp. 257-272, Copyright 2001, with permission from Elsevier.)

developments include making long oligonucleotides for higher sensitivity (e.g., the 60 mer microarray show a five- to tenfold increase in sensitivity over shorter ologonucleotides) (Constans, 2003b). Moreover, rapid, low-cost manufacturing techniques, such as inkjet-style processes, promise to reduce the cost of microarrays. In addition, new microarrays allow cells to be viewed under thousands of different growth conditions and physiological states (Sedlak, 2003). Recent developments include attaching PCR amplification fragments on the slide, rather than just the small nucleotides (Willis, 2003). GeneSifter.Net, a new simple software package for microarray data analysis, is now accessible through the Web, providing an easier way to manage data and obviating the need for licenses (Constans, 2003a).

DNA microarray technology has been used for the identification of organisms in the environment and clinical or food samples based on the presence of unique sequences (e.g., 16S rRNA, 23S rRNA, and key functional genes for the transgenic organisms) (Troesch et al., 1999). Robust and reliable target amplification methods have been developed to enable transcript profiling from submilligram amounts of plant tissue employing 3' cDNA tag amplification and subsequent PCR reaction. A twofold expression difference could be distinguished with 99 percent confidence using only 0.1 µg total plant RNA tissue (Hertzberg et al., 2001).

For field applications, a portable system for microbial sample preparation and oligonucleotide microarray analysis was reported (Bavykin et al., 2001). This portable system contained three components:

1. Universal silica minicolumn for successive DNA and RNA isolation, fractionation, fragmentation, fluorescent labeling, and removal of excess free label and short oligonucleotides
2. Microarrays of immobilized oligonucleotide probes for some gene sequences (e.g., 16S rRNA) identification
3. A portable, battery-operated device for imaging the hybridization of fluorescently labeled RNA fragments on the array

Beginning with a whole cell, it takes approximately 25 min to obtain labeled DNA and RNA samples, and an additional 25 min to hybridize and acquire the microarray image using a stationary image analysis system or the portable imagery (Bavykin et al., 2001). This portable system can be modified and utilized for field detection of GMOs.

A challenge to microarray technology is standardization to ensure that data collected from different microarray platforms can be accurately compared (Constans, 2003b). Many modifications to make arrays flexible, or to use them for high- or low-density applications to cut cost, or to produce a chip with increased density that can be baked into smaller or larger pieces depending on the application have been made by companies such as Affymetrix. Other companies produce modular chips, such as SensiChip from QIAGEN, which contain six arrays separated by microfluid hybridization chambers, allowing users to run six different experiments on a single bar for low-density applications. Moreover, the density of the bar can be increased further by dividing the content on several microarrays, allowing for as few as one experiment per bar, thereby cutting cost (Constans, 2003b). Because different expressions can be due to differential expression of specific splice variants of a given gene, and not necessarily the gene's overall expression, probes that reflect mRNA variations are being manufactured (Shoemaker and Linsley, 2002).

The hindrances to utilization of the microarray technique on a wide scale in the biosafety arenas include the perceived expense and availability of the technology. However, robotics for handling large numbers of clones and spotting arrays are now available at moderate cost, and are not difficult to operate. Although the testing facility or the research laboratory must overcome technical difficulties, it is feasible to establish a 5000-clone microarray resource within 12 months of commencing a project, and within a reasonable budget. A typical basic experiment involving 50 or so pairwise hybridizations would cost between $5,000 and $10,000 in reagents. An investment of about $1 million can fund the creation of a DNA microarray facility (Ahmed, 2002b; Gibson, 2002).

DNA-BASED BIOSENSORS

An affinity biosensor is a device incorporating immobilized receptor molecules on a sensor surface, which can reversibly detect—nondestructively—receptor-ligand interactions with a high differential selectivity. A stoichiometric binding event occurs in the sensors, and the associated physicochemical changes are detected by a transducer

(Marco and Barcelò, 1996). The biorecognition elements currently employed are receptors, antibodies, and nucleic acids (Sedlak, 2003). Nucleic acids were used recently as bioreceptors in biosensors and biochips (Vo-Dinh and Cullum, 2000). These biosensors are based on the immobilization of a DNA probe, and on the monitoring and recording of the variation of a transducer signal when the complementary target in solution interacts with the probe forming a stable complex (Marrazza et al., 1999). For GMOs, two types of real-time biosensing transducers are used at present: piezoelectric, in the form of quartz crystal microbalance (QCM), and surface plasmon resonance (SPR) (Mariotti et al., 2002; Minunni et al., 2001).

In piezoelectric detection, an electric field is applied to a material, causing a minute mass change that can be detected electronically. For example, in the QCM device an electronic circuit drives an AT-cut quartz crystal that is sandwiched between two electrodes producing resonance frequency. This signal is tracked with a frequent counter. Changes in the resonance frequency are then related to changes in the mass of the crystal, thus allowing the QCM to function as a mass detector (Figure 8.8). A typical sensor response of QCM during DNA hybridization in solution is presented in Figure 8.9. Here, a nucleic acid probe is immobilized on gold (Au). Two buffers are used. After buffer (#1) is added to equilibrate the surface, the probe is exposed to the target analyte—a complementary DNA sequence made of synthetic oligos—which binds to the probe. The hybridization reaction causes a mass increase on the crystal surface followed by a decrease in the resonance frequency of the crystal. The unbound analyte is removed by washing the crystal in the buffer, and the resonance frequency is recorded (#2). The difference between (#2) and (#1) is the frequency shift (f in Hz) caused by the analyte. A calibration curve is then constructed by plotting f versus analyte concentration (Figure 8.10). The unknown concentration of target analyte can thus be determined (Minunni, 2003). Figure 8.11 shows a piezoelectric sensor (Model QCA917; Seiko EG&G).

SPR is based on transferring photons (light energy) to a plasmon (a group of oscillating electrons on a metal surface). Light is coupled to the plasmon surface using either a prism or a metal grating. A thin film of differing dielectric (such as Au) is immobilized on the opposite surface of the metal (Löfas and Johnsson, 1990). Chemical changes in the film and the light reflected from the prism interact at a resonant

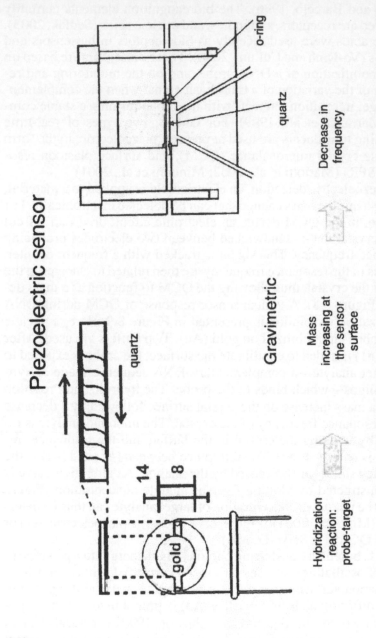

FIGURE 8.8. Schematic of a piezoelectric sensor. 10 MZ AT-cut quartz crystal with gold electrodes and methacrylate measurement cell for analysis in liquid phase. (*Source:* From Minnuni, 2003.)

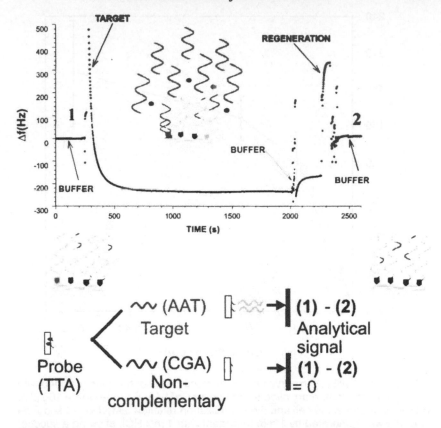

FIGURE 8.9. A typical sensor response of piezoelectric sensing during DNA hybridization experiments. The testing sample (target DNA) is injected over the sensing surface with the immobilized probe and the hybridization interaction occurs. A decrease in frequency is observed. To remove unbound material, the surface is washed with buffer. The surface is regenerated, dissociating the affinity DNA bound with acid. The analytical signal is the difference between the value recorded before the sample injection (baseline) and the one ordered after the buffer washing. (*Source:* From Minunni, 2003.)

angle that depends on the molecular composition of the surface. Chemical changes in that film then modulate the reflected light from the prism. The reflected light is monitored and the resonant condition is indicated by a decrease in intensity of the reflected light as a function of the incident angle. The angle at which the minimum occurs is referred to as the SPR angle (Figure 8.12).

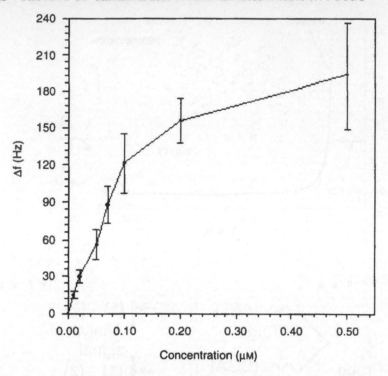

FIGURE 8.10. Calibration curve for a piezoelectric sensor that was obtained with the 25-mer complementary oligonucleotide for CaMV 35S probe where 100 μl of sample is added in the cell and the hybridization reaction lasted for 10 sec. The ss-probe was generated by 1 min treatment with 1 mM HCl, allowing a successive hybridization reaction to be monitored. (*Source:* From Minunni, 2003.)

The hybridization interaction between nucleic acids is an affinity interaction. When a synthetic oligo is immobilized on the sensor surface, and the complementary strand is free in solution, the hybrid formation between the two nucleic acids strands can be monitored in real time, as shown in Figure 8.13. After a first addition of buffer (#1) to equilibrate the surface, the probe is exposed to the target analyte (a complementary DNA sequence of synthetic nucleotides that binds the probe). The hybridization reaction causes a change in the refractive index, which influences the resonant angle recorded as a shift (#2). The difference between (#2) and (#1) is the resonance shift that is expressed in arbitrary resonance units (RU). Plotting RU versus the

Optical sensor: Biacore X (Biacore AB)	Piezoelectric sensor: Model QCA917 (Seiko EG&G)
 *Sensor Chip*	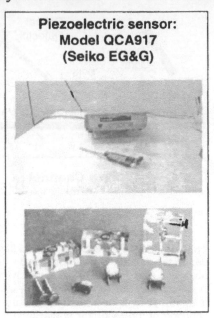

FIGURE 8.11. Instrumentation used for sensing. An optical SPR sensor, Biacore X (Biacore BA), and a piezoelectric sensor, Model QCA 917 (Seiko EG&G). (*Source:* From Minunni, 2003.)

analyte concentration gives rise to a calibration curve that can be used to estimate the concentration of target analyte. An instrument based on this technology—the Biacore (Biacore AB)—is shown in Figure 8.11. The QCM and SPR systems, although sensitive to mass and refractive index changes, respectively, are prone to errors due to unspecific adsorption, since unspecific binding can still generate a signal (Minunni, 2003).

For the development of DNA-based sensors, a probe is immobilized on the Au-bound surface, either directly (i.e., without the presence of a linking lawyer), or indirectly by anchoring via streptavidin protein, requiring use of a biotinylated probe because of the very high affinity between avidin and biotin, thus creating a stable DNA surface for interaction with complementary target DNA (Tombelli et al., 2000).

For GMO detection, QCM and SPR biosensors have been used by immobilizing specific 25-mer oligonucleotide probes using different

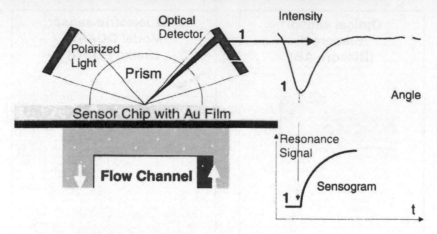

FIGURE 8.12. SPR sensing. In a condition of total internal reflection, the evanescent wave generates plasmons, and a dip in the intensity of the reflected light is observed. The incident angle at which the minimum is recorded is called the *resonant angle*. The resonant angle—influenced by the refractive index in the medium in which the interaction occurs—is monitored. The graph displaying the variation of the resonant angle versus time is the sensogram. (*Source:* From Minunni, 2003.)

sensing surfaces (e.g., a screen-printed electrode [or chip], the piezoelectric crystal, and the Biacore sensor [M5] dextran-coated chip). These sequences are complementary to the sequences of the CaMV 35S-promoter and the NOS terminator. The base sequence of the 5'-biotinylated probes (25 mer), the complementary nucleotides (25 mer), and the noncomplementary oligos (23 mer) are:

> Probe CaMV 35S 5'biotin-
>   GGCCATCGTTGAAGATGCCTCTGCC 3'
> Probe NOS 5'biotin- AATGATTAATTGCGGGACTCTAATC 3'
> Target 35S 5' GGCAGAGGCATTCAACGATGGCC 3'
> Target NOS 5' GATTAGAGTCCCGCAATTAATCATT 3'
> Noncomplementary strand 5'
>   TGCCCACACCGACGGCGCCCACC 3'

Biotinylation of the probes is required for the piezoelectric and SPR sensors, but not for the electrochemical one. The PCR probe must first be denatured to produce a single-stranded (ss) DNA capable of hybridization. For QCM sensing, denaturation at 95°C for 5 min

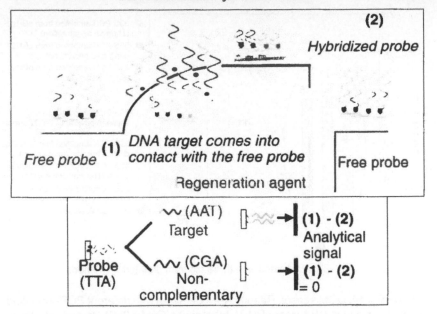

FIGURE 8.13. A sensogram displaying the affinity interaction between the immobilized probe and the complementary target in solution. The increase in the sensor signal can be observed while the interaction occurs. The system operates in flow. After the surface is washed, the hybridization signal is taken as the difference between the values recorded before the interaction (baseline) and after the washing. (*Source:* From Minunni, 2003.)

followed by incubation on ice was reported to be adequate (Tombelli et al., 2000). For SPR sensing using Biacore X, the thermal treatment did not allow adequate amount of ss DNA to reach the sensor surface and react with the probe. However, when magnetic beads were used, the problem was overcome. Figure 8.14 shows QCM signals for three different denaturing methods: thermal, enzymatic, and magnetic using CaMV 35S PCR-amplified plasmid DNA. Signals obtained from GMO samples for different foods (e.g., dietetic snacks and drinks, and certified reference material) were treated only thermally. The system was optimized using synthetic complementary oligos (25 mer), and the specificity of the system (which relies on the immobilized probe sequence) was tested with a noncomplementary oligo (23 mer). The hybridization study was performed also with DNA samples from

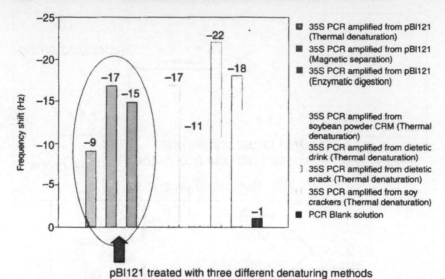

pBI121 treated with three different denaturing methods

FIGURE 8.14. QCM sensor. Results were obtained with different PCR-amplified samples: plasmid DNA from pBI121 (containing CaMV 35S) treated with three different denaturation methods (thermal, enzymatic, and magnetic particles); transgenic CRM, and dietetic snacks and light drinks (also containing CaMV 35S). (*Source:* From Minunni, 2003.)

CRM soybean powder (EC, 2001) containing 2 percent GMO and amplified by PCR. Nonamplified genomic or plasmid DNA were also used. The amplified CaMV 35S resulted in a fragment 195 bp long (Minnuni, 2003).

Biacore X is priced at $112,000. The chip employed for SPR analysis (e.g., dextran modified) costs about $200, with one chip capable of performing 100 analyses of PCR amplified samples. A system based on piezo sensing is, however, much cheaper, averaging about $13,000 (including software). The sensing element (i.e., piezoelectric crystal) costs about $25, allowing up to 25 analyses per surface, and doubling the number of reactions with both surfaces used. Reagents for both methods cost nearly the same (about $70 for biotinylated probes immobilized on the transducer surface). About 35 probes are needed for the analysis, making a probe cost per chip in the range of $1. Synthetic oligonucleotides are needed to generate a calibration curve (at a cost of $3 per chip).

Such affinity biosensor systems are attractive for DNA sensing because their versatility is often associated with probability and with the absence of labeling. In addition, many analyses can be performed on the same sensor surface with the possibility of reuse of the QCM and SPR devices up to 30 times and more than 100 times, respectively (Marrazza et al., 1999). Future prospects in this rapidly developing area will result—within few years—in new equipment and formats that will enhance the detection sensitivity of these methods.

Based on affinity biosensing, GeneScan Europe recently introduced a test kit for the detection of GMOs in food products that allows a multiplex PCR for the specific detection of DNA sequences from plant species and GM traits using a biosensor chip and a biochip reader (GeneScan Europe, 2001). The procedure begins with the isolation and purification of DNA from the sample. Specific DNA sequences from plant species and cereals (e.g., corn, canola, soy, rice) and GM traits (e.g., CaMV 35S promoter, NOS terminator, *bar* gene) are then amplified by two separate multiplex PCR reactions. The products of both reactions are mixed, and ss DNA is created by digestion with an exonuclease. After mixing with a hybridization buffer, the sample is spread on the chip, and amplified sequences that will hybridize with cDNA probes covalently bound in the chip are stained with the fluorescent dye Cys and analyzed by a biochip reader, such as Biodetect 654. The detection limit for the kit is in the range of 250 copies of each of the target DNA sequences in the PCR (GeneScan Europe, 2001).

Developments in nanotechnology have allowed biochemically induced surface stress to directly and specifically transduce molecular recognition into a nanomechanical response using a cantilever array (Fritz et al., 2000). Cantilevers are microfabricated silicon devices that are gold coated, 1 μm thick, 500 μm long, and 100 μm wide. The upper surface can be coated with a DNA oligonucleotide or antibody molecule. Specific binding of the complimentary molecule causes a difference in surface stress that leads to the cantilever, which in turn can be detected by light beams whose angle of reflection is altered. DNA probes can be dissociated, regenerating the device and allowing it to be reused several times. Based on this technology a DNA chip can be developed that will produce parallelization in integrated devices, using an array of more than 1,000 cantilevers at a low cost without the need to amplify DNA by PCR. Samples can be measured directly without amplification (Morrow, 2002).

CONCLUSIONS

Hybridization probes are qualitative basic research tools for the detection of GMOs in foods. They are not practical for routine testing because of their slowness and low throughput. Quantitative competitive and real-time PCR tests are sensitive and reliable tests for the quantitation of GMO content in foods, although the latter assay is more suited for routine testing because of the speed and large dynamic range of amplification, thus allowing for higher throughput analysis than conventional PCR assays. Microarray technology, based on measuring changed expression as a result of introduced gene(s), is fast developing as efforts to increase the sensitivity and specificity of its detection ability has been accelerated, and has been applied for identifying GMOs in food. Microarrays are automated, provide high throughput, and allow for parallel analysis of thousand of genes. Drawbacks include high initial expenses for constructing a DNA microarray facility and the need for specialized software to analyze the generated data, making it not easily applicable for routine testing in small facilities and unattractive for inexperienced operators. DNA sensing is also an emerging technology based on hybridization between an immobilized DNA probe and a molecular target consisting of a probe complementary sequence in solution. Surface plasmon resonance and piezoelectric sensing have been employed as transduction sensing devices. These sensors are able to monitor in real time, and are suitable for field applications to GMO detection. Future improvements in method sensitivity promise to increase their wider applicability for GMO testing.

REFERENCES

Ahmed, F.E. (1995). Application of molecular biology to biomedicine and toxicology. *J. Environ. Sci. Health* C11:1-51.

Ahmed, F.E. (1998). Molecular methods for detection of mutations. *J. Environ. Sci. Health* C16:47-80.

Ahmed, F.E. (1999). Safety standards for environmental contaminants in foods. In *Environmental Contaminants in Foods* (pp. 500-570), C. Moffat and K.J. Whittle, eds. Sheffield Academic Press, United Kingdom.

Ahmed, F.E. (2000). Molecular markers for cancer detection. *J. Environ. Sci. Health* C18:75-126.

Ahmed, F.E. (2002a). Detection of genetically modified organisms in foods. *Trends Biotechnol.* 20:215-223.

Ahmed, F.E. (2002b). Molecular techniques for studying gene expression in carcinogenesis. *J. Environ. Sci. Health* C20:71-116.

Bavykin, S.G., Akowski, J.P., Zakhariev, V.M., Barsky, V.E., Perov, A.N., and Mirzabekov, A.D. (2001). Portable system for microbial sample preparation and oligonucleotide microarray analysis. *Appl. Environ. Microbiol.* 67:922-928.

Brodmann, P.D., Ilg, E.C., Berthoud, H., and Herman, A. (2002). Real-time quantitative polymerase chain reaction methods for four genetically modified maize varieties and maize DNA content in food. *J. AOAC Int.* 85:646-653.

Constans, A. (2003a). Array analyses online. *The Scientist* 17:41.

Constans, A. (2003b). Challenges and concerns with microarrays. *The Scientist* 17: 35-36.

Constans, A. (2003c). Microarray instrumentation. *The Scientist* 17:37-38.

Cortese, J.D. (2000). The array of today. *The Scientist* 14:25-28.

Denhardt, D.T. (1966). A membrane-filter technique for the detection of complementary DNA. *Biochem. Biophys. Res. Commun.* 23:641-653.

Doblhoff-Dier, O., Bachmayer, H., Bennett, A. Brunius, G., Bürki, K., Cantley, M., Collins, C., Crooy, P., Elmquist, A., and Frontali-Botti, C. (1999). Safe biotechnology: Values in risk assessment for the environmental application of microarrays. *Trends Biochem.* 17:307-311.

Drennan, J.L., Westra, A.A.G., Slack, S.A., Delserone, L.M., and Collmer, A. (1993). Comparison of a DNA hybridization probe and ELISA for the detection of *Clavibacter michiganensis* subsp. *sepodonicus* in field-grown potatoes. *Plant Dis.* 77:1243-1247.

Ermolaeva, O., Rastogi, M., Pruitt, K.D., Schler, G.D., Bittner, M.L., Chen, Y., Simon, R., Meltzer, P., Trent, J.M., and Boguski, M.S. (1998). Data management and analysis for gene expression arrays. *Nat. Genet.* 20:19-23.

European Commission (EC) (1997). Council Regulation No 285/97, 27 January 1997, concerning Novel Foods and Novel Food Ingredients. *Official J. European Communities*, No. L43, 1-5.

European Commission (EC) (2001). Certified Reference Material IRMM-413. Dried Maize Powder Containing Genetically Modified MON 810 Maize. IRMM Unit for Reference Materials. Resticseweg, B-2440, Geel, Belgium, January.

European Commission (EC) (2002). Certified Reference Material IRMM-410S Dried Soya Bean Powder Containing Genetically Modified Roundup Ready Soya Beans. IRMM Unit for Reference Materials. Resticseweg, B-2440, Geel, Belgium, March.

Fritz, J., Baller, M.K., and Lang, H.P. (2000). translating biomolecular recognition into nanomechanics. *Science* 288:316-318.

Gasch, A., Wilborn, F., Schft, P., and Berghof, K. (1997). Detection of genetically modified organisms with the polymerase chain reaction: Potential problems with the food matrix. In *Foods Produced by Means of Genetic Engineering* (pp. 90-97), Second Edition, F.A. Schreiber and K.W. Bögl, eds. BgVV-hafte.01-1997.

Gasser, C.S. and Fraley, R. (1989). Genetically engineering plants for crop improvement. *Science* 244:1293-1299.

GeneScan Europe (2000). Testing for genetically modified organisms. Practical applications of PCR. A presentation of USDA/AEIC Grain Biotechnology Detection Methods Validation Workshop, February 24-25, 2000, Kansas City, MO.

GeneScan Europe (2001). *GMO Chip: Test Kit for the Detection of GMOs in Food Products*, Cat. No. 5321300105, Bermen, Germany.

Gibson, G. (2002). Microarrays in ecology and evolution: A preview. *Molec. Ecol.* 11:17-24.

Graves, D.J. (1999). Powerful tools for genetic analysis come of age. *Trends Biotechnol.* 17:127-134.

Hacia, J.G., Makalowski, W., Edgemon, K., Erdos, M.R., Robbins, C.M., Fodor, S.P.A., Brody, L.C., and Collins, F.S. (1998). Evolutionary sequence comparisons using high-density oligonucleotide array. *Nature Genet.* 18:155-158.

Hardegger, M., Brodmann, P., and Hermann, A. (1999). Quantitation detection of the 35S promoter and the NOS terminator using quantitative competitive PCR. *Eur. Food Res. Technol.* 28:83-87.

Hemmer, W. (1997). *Foods Derived from Genetically Modified Organisms and Detection Methods*. BATS-report 2/97, Bale, Switzerland.

Hertzberg, M., Sievertzon, M., Aspeborg, H., Nilsson, P., Sandberg, G., and Lundeberg, J. (2001). cDNA microarray analysis of small plant tissue sample using a cDNA tag target amplification protocol. *Plant J.* 25:585-591.

Hill, M., Melanson, D., and Wright, M. (1999). Identification of the APH IV gene from protoplast-derived maize plants by a simple nonradioactive DNA hybridization method. *In Vitro Cell. Dev. Biol.-Plant* 35:154-155.

Hübner, P., Studer, E., and Lüthy, J. (1999a). Quantitation of genetically modified organisms in food. *Nature Biotechnol.* 17:1137-1138.

Hübner, P., Studer, E., and Lüthy, J. (1999b). Quantitative competitive PCR for the detection of genetically modified organisms in food. *Food Control* 10:353-358.

Hübner, P., Waiblinger, H.-U., Pietsch, K., and Brodmann, P. (2001). Validation of PCR methods for the quantification of genetically modified plants in food. *J. AOAC Int.* 84:1855-1854.

Hupfer, C., Hotzel, H., Sachse, K., and Engel, K.-H. (1998). Detection of the genetic modification in heat-treated products of Bt maize by polymerase chain reaction. *Z. Lebensum. Unters. Forsch* 206:203-207.

Ikuta, S., Takag, K., Wallace, R.B., and Itakura, K. (1987). Dissociation kinetics of 19 base-paired oligonucleotide-DNA duplexes containing different single mismatched base pair. *Nucleic Acids Res.* 15:797-802.

Jankiewicz, A., Broll, H., and Zagon, J. (1999). The official method for the detection of genetically modified soybeans (German Food Act LMBG § 35): A semi-quantitative study of sensitivity limits with glyphosate-tolerant soybeans (Roundup Ready) and insect resistant Bt maize (Maximizer). *Eur. Food Res. Technol.* 209:77-82.

Johnson, D.A., Gantsch, J.W., Sportsman, J.R., and Elder, J.H. (1984). Improved technique utilizing nonfat dry milk for analysis of proteins and nucleic acids transferred to nitrocellulose. *Gene Anal. Tech.* 1:3-12.

Jordan, B.R. (1998). Large-scale expression measurement by hybridization methods: From high-density membranes to "DNA chips." *J. Biochem.* 124:251-258.

Köppel, E., Stadler, M., Lüthy, J., and Hübner, P.H. (1997). Sensitive Nachweismethode für die gentechnisch veränderte Sojabohne "Roundup Ready." *Mitt. Gebiete Lebensm. Hyg.* 88:164-175.

Kuipers, O.P. (1999). Genomics for food biotechnology: Prospects for the use of high throughput technologies for the improvement of food in microorganisms. *Curr. Opinion Biotechnol.* 10:511-516.

Löfas, S. and Johnsson, B. (1990). A novel food hydrogel matrix on gold surfaces in surface plasmon resonance sensor for fast and efficient covalent immobilization of ligands. *J. Chem. Soc. Chemical Comm.* 21:1526-1527.

Lüthy, J. (1999). Detection strategies for food authenticity and genetically modified foods. *Food Control* 10:359-361.

Marco, M.-P. and Barceló, D. (1996). Environmental applications of analytical biosensors. *Measure Sci. Technol.* 7:1547-1562.

Mariotti, E., Minunni, M., and Mascini, M. (2002). Surface plasmon resonance biosensor for genetically modified organisms detection. *Anal. Chimica Acta* 453: 165-172.

Marrazza, G., Chianella, I., and Mascini, M. (1999). Disposable DNA electrochemical sensor for hybridization detection. *Biosensors Bioelectronics* 14:43-51.

Meyer, R. (1995) Nachweis gentechnisch veränderter Pflanzen mittels PCR am Beispiel der Flavr Savr-Tomate. *Z. Lebensum. Unters. Forsch.* 201:583-586.

Meyer, R. (1999). Development and application of DNA analytical methods for the setection of GMOs in food. *Food Control* 10:391-399.

Minunni, E. (2003). Biosensors based on nucleic acid interaction. *Spectroscopy* 17: 613-635. Figures reprinted with permission of IOS Press.

Minunni, M., Tombelli, S., Mariotti, E., Mascini, M., and Mascini, M. (2001). Biosensors as new analytical tool for detection of genetically modified organisms (GMOs). *Fres. J. Anal. Chem.* 369:589-593.

Morrow, J. Jr. (2002) Nanotechnology: Application to biotechnology. *American Genomic/Proteomic Technol.* (November/December):26-29.

Nuwaysir, E.F., Bittner, M., Trent, J., Barrett, J.C., and Afshari, C.A. (1999). Microarrays and toxicology: The advent of toxigenomics. *Molec. Carcino.* 24: 153-159.

Osborn, J. (2000). A review of radioactive and non-radioactive based techniques used in life science applications–Part I: Blotting techniques. *Life Science News* (Amersham Biosciences) 6:1-4.

Pease, A.C., Solas, D., Sullivan, E.J., Coronin, M.T., Holmes, C.P., and Fodor, S.P.A. (1994). Light-generated oligonucleotide arrays for rapid DNA sequence analysis. *Proc. Natl. Acad. Sci. USA* 91:5022-5026.

Pietsch, K., Waiblinger, H.U., Brodmann, P., and Wurz, A. (1997). Screening-verfahren zur Identifizierung "gentechnisch veränderter" pflanzlicher Lebensmittel. *Dtsch. Lebensum. Rundsch* 93:35-38.

Raeymackers, L. (1993). Quantitative PCR: Theoretical considerations with practical applications. *Anal. Biochem.* 214:582-585.

Rivin, C.J., Cullis, C.A., and Walbot, V. (1986). Evaluating quantitative variation in the genome of *Zea mays*. *Genetics* 113:1009-1019.

Ross, R., Ross, X.-L., Rueger, R.B., Laengin, T., and Reske-Kunz, A.B. (1999). Nonradioactive detection of differentially expressed genes using complex RNA or DNA hybridization probes. *BioTechniques* 26:150-155.

Ruan, Y., Gilmore, J., and Coner, T. (1998). Towards *Arabidopsis* genome analysis: Monitoring expression profiles of 1400 genes using cDNA microarrays. *Plant J.* 15:821-833.

Sambrook, J. and Russel, D. (2000). *Molecular Cloning: A Laboratory Manual*, Third Edition. Cold Spring Harbor Laboratory Press. Cold Spring Harbor, NY.

Schena, M., Shalon, D., Davis, R.W., and Brown, P.O. (1995). Quantitative monitoring of gene expression patterns with a complementary DNA microarray. *Science* 270:467-470.

Schreiber, G.A. (1999). Challenges for methods to detect genetically modified DNA in foods. *Food Control* 10:351-352.

Schummer, M., Ng., W.-I., Nelson, P.S., Bumgarner, R.E., and Hood, L. (1997). Inexpensive handheld device for the construction of high-density nucleic acid arrays. *BioTechniques* 23:1087-1092.

Sedlak, B.J. (2003). Next-generation microarray technologies. *Genet. Engineer. News* 23:20, 59.

Shoemaker, D.D. and Linsley, P.S. (2002). Recent developments in DNA microarrays. *Curr. Opin. Microbiol.* 5:334-337.

Slack, S.A., Drennan, J.L., and Westra, A.A.G. (1996). Comparison of PCR, ELISA, and hybridization for the detection of *Clavibacter michiganensis* subsp. *sepedonicus* in field grown potatoes. *Plant Dis.* 80:519-524.

Southern, E.M. (1975). Detection of specific sequences among DNA fragments separated by gel electrophoresis. *J. Mol. Biol.* 98:503-517.

Studer, E., Rhyner, C., Lüthy, J., and Hübner, P. (1998). Quantitative competitive PCR for the detection of genetically modified soybean and maize. *Z. Lebensum. Unters. Forsch.* A207:207-213.

Stull, D. (2001). A feat of fluorescence. *The Scientist* 15:20-21.

Swiss Food Manual (Schweizerisches Lebensmittelbuch) (1998). Bundesamt für Gesundheit (Ed.) Kapitel 52B: Molekularbiologische Methoden. Eidgenössische Drucksachen- und Materialzentrale, Bern, Switzerland.

Sykes, P.J., Neoh, S.H., Brisco, M.J., Hughes, E., Condon, J., and Morley, A.A. (1998). Quantitation of targets for PCR by use of limiting dilution. In *The PCR Technique: Quantitative PCR* (pp. 81-93), J.W. Larrick, ed. Eaton Publishing Co., Nattick, MA.

Tengel, C., Scübler, P., Setzke, E., Balles, J., and Sprenger-Haussels, M. (2001). PCR-based detection of genetically modified soybean and maize in raw and highly processed food stuff. *BioTechniques* 31:426-429.

Tombelli, S., Mascini, M., Sacco, C., and Turner, A.P.F. (2000). A DNA piezoelectric biosensor assay coupled with a polymerase chain reaction for bacterial toxicity determination in environmental samples. *Anal. Chimica Acta* 418:1-9.

Troesch, A., Nguyen, H., Miyada, C.G., Desvarenne, S, Gingeras, T.R., Kaplan, P.M., Cros, P., and Mabilat, C. (1999). *Mycobacterium* species identification and rifampin resistance testing with high-density DNA probe arrays. *J. Clin. Microbiol.* 37:49-55.

Van Hal, N.L.W., Vorst, O., van Houwelingen, A.M.M.L., Kok, E.J., Peijumen-burg, A., Aharoni, A., van Tunen, A.J., and Keijer, J. (2000). The application of DNA microarrays in gene expression analysis. *J. Biotechnol.* 78:271-280.

Van Hoef, A.M.A., Kok, E.J., Bouw, E., Kuiper, H.A., and Keijer, J. (1998). Development and application of a selective detection method for genetically modified soy-derived products. *Food Addit. Contamin.* 15:767-774.

Vo-Dinh, T. and Cullum, B. (2000) Biosensors and biochips: Advances in biological and medical diagnostics. *Fresenius J. Anal. Chem.* 366: 540-551.

Vollenhofer, S., Bul, K., Schmidt, J., and Kroath, H. (1999). Genetically modified organisms in food—Screening and specific detection by polymerase chain reaction. *J. Agric. Food Chem.* 47:5038-5043.

Willis, R.C. (2003). Monitoring microbes. *Modern Drug Discovery* (January):16-21.

Wilson, W.J., Strout, C.L., DeSantis, T.Z., Stilwell, J.L., Carrano, A.V., and Andersen, G.L. (2002). Sequence-specific identification of 18 pathogenic microorganisms using microarray technology. *Mol. Cell. Probes* 16:119-127.

Wurz, A., Bluth, A., Zeltz, P., Pfeifer, C., and Willmund, R. (1999). Quantitative analysis of genetically modified organisms (GMO) in processed food by PCR-based methods. *Food Control* 10:385-389.

Ye, R.W., Wang, T., Bedzyk, L., and Croker, K.M. (2001). Application of DNA microarrays in microbial systems. *J. Microbiol.* 47:257-272.

Zhou, J. and Thompson, D.K. (2002). Challenges in applying microarrays to environmental studies. *Curr. Opinion Biotechnol.* 13:204-207.

Chapter 9

Near-Infrared Spectroscopic Methods

Sylvie A. Roussel
Robert P. Cogdill

This chapter presents an alternative technique to the DNA and protein-based methods for genetically modified organism (GMO) screening. First, some reasons are offered for the common reluctance of the commodity food industry to more often utilize DNA and protein-based methods for GMO screening, which suggest a need for an appropriate alternative. Second, an introduction to near-infrared (NIR) spectroscopy is presented. Third, a case study for the detection of Roundup Ready (RR) soybeans is detailed. Finally, the conditions of application and the future developments of NIR usage are discussed.

INTRODUCTION: THE LIMITATIONS OF APPLYING CURRENT GMO SCREENING METHODS

Although a number of protein-based and DNA-based GMO screens—hereafter referred to as "direct methods"—have been developed with high degrees of accuracy and specificity, some sectors of the food industry are still lacking a suitable GMO screening tool. For a variety of reasons, many producers and handlers of bulk com-

The research work concerning the discrimination of Roundup Ready soybeans with NIR spectroscopy was carried out at the Grain Quality Laboratory (GQL), Department of Agricultural and Biosystems Engineering of Iowa State University (ISU), Ames, Iowa (http://www.iowagrain.org). The authors acknowledge their co-workers, Charles R. Hurburgh Jr., head of GQL; Glen R. Rippke, Laboratory Manager; and Connie L. Hardy, Iowa State University Center for Crops Utilization Research. The ISU Agronomy department is also thanked for providing data from two spectrometers at the agronomy farm of Iowa State University.

modities, especially grain and feed, have resisted developing comprehensive GMO screening plans with direct methods as their only option for detection. In particular, their resistance stems from the following realities of applying direct methods for GMO screening in bulk commodities:

1. *They incur relatively large per-sample costs:* Many sectors of the bulk food industry are faced with the competing demands of narrow per-unit profit margins, and have virtually no control over the price received for goods sold. Because little room exists to absorb new costs, and because it is difficult to pass new costs on to consumers, bulk food handlers are reluctant to adopt any new practice that will increase costs.

For example, a country grain elevator that generally handles 400 to 900 bushel loads of grain *nets* approximately $0.012 to $0.024 per bushel of soybeans handled or $4.80 to $21.60 per incoming lot. Thus, a GMO screen using $2.90/sample test kit (GIPSA Certified RUR strips, 2002, Strategic Diagnostics, Inc., Newark, DE) can devour a significant portion of a country elevator's profit if every lot is to be screened. Furthermore, the total cost of a GMO screen may include additional sample handling and grinding costs, as well as the cost of acquiring qualified labor, all of which may add considerably to the price of the test kit. This is important because small country grain elevators are often the final point in the grain handling system before which significant mixing occurs.

2. *They are time-consuming:* Whether it is because of product freshness concerns, labor costs, or harvest time constraints, the bulk food industry has allotted very little time (per unit) for product analysis (Figure 9.1 shows grain trucks lining up for inspection which consumes considerable time). During a study conducted at a country elevator to determine the frequency distribution of grain delivery during harvest (Hurburgh, 1999b), it was found (Figure 9.2) that during the peak harvest days, the elevator received more than 160 loads per day, leaving less than 10 minutes to process each load. Thus, it would be impossible to perform an additional GMO screen, using even the fastest of direct methods, without significantly sacrificing productivity.

3. *A sampling issue exists:* It is difficult to gather a small, yet still representative sample: For direct methods of GMO detection, which invariably require a relatively small sample (on the order of 25 g), the problem of sampling error can be quite large (Hurburgh, 1999a).

FIGURE 9.1. Trucks of grain line up for inspection and unloading at an elevator near Charleston, South Carolina (*Source:* USDA Photo by Ken Hammond, <http://www.usda.gov/oc/photo/opcservj.htm>.

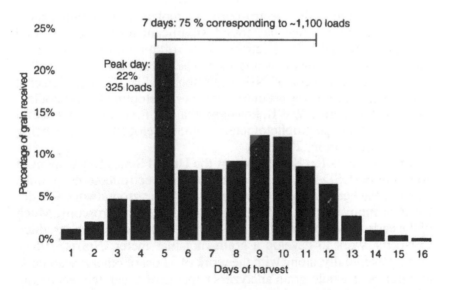

FIGURE 9.2. Daily soybean-receiving patterns at a country elevator in the U.S. Midwest

Indeed, finding a small concentration of GM seeds in a truckload of grain can be akin to searching for a needle in a haystack, particularly when the needle looks just like the hay. The problem is exacerbated by mixed or blended lots, which is often the norm for the latter stages of bulk commodity supply chains. Although the problem of generating a representative sample can be reduced by grinding and thoroughly mixing the sample (USDA, 1995), all of this adds to the time burden imposed by the screening method.

It is clear that the commodity food industry is in need of a more feasible solution to GMO detection if comprehensive screening is to be implemented at all levels of the supply chain. Because of a number of efficiencies in application, near-infrared spectroscopy has been suggested as a potentially attractive alternative to direct methods for the detection of GMO in bulk commodities.

INTRODUCTION TO NEAR-INFRARED SPECTROSCOPY

History

Near-infrared (NIR) spectroscopy is a molecular vibrational spectroscopy technique that has found significant application in food analysis. Although NIR spectroscopy is still a relatively novel technique in terms of growth, its origins can be traced back more than 200 years to the discovery of NIR radiation by Sir William Herschel (Herschel, 1800). It wasn't until after the development of photoelectric detectors during WWII, however, that the first practically useful NIR recording spectrophotometers were developed, the first being the Cary 14 (Barton, 2002).

Shortly thereafter, Karl Norris of the United States Department of Agriculture (USDA), armed with a Cary 14 spectrophotometer and a Hewlett-Packard calculator, began laying the groundwork for decades of innovation in food analysis using NIR spectroscopy. Much of the early development of the NIR analyzer industry can be credited to Norris' work to develop a grain-moisture meter (Norris, 1964), leading eventually, along with the work of a host of other researchers, to families of whole-grain analyzers capable of accurately analyzing *intact* samples of grain for a variety of quality attributes (Shenk and Westerhaus, 1993; Williams and Sobering, 1993; AACC, 1999).

At present, NIR analyzers are used commonly at commodity food handling facilities throughout the world (Figure 9.3). Their combination of analytical power, rapidity and ease of application, nondestructive nature, and low cost have made them the standard for a growing list of food analysis tasks.

Beyond whole grain, NIR spectroscopy has been used to analyze other bulk products, such as raw feed components, forage (Sinnaeve et al., 1994), and manure (Millmier et al, 2000; Reeves, 2001). NIR spectroscopy has also had a positive impact on the quality analysis of most agricultural and food products (Williams and Norris, 2001), such as meat (Naes and Hildrum, 1997), dairy (Sasic and Ozaki, 2000), or beverage (Rodriguez-Saona et al., 2001), and on process control (Berntsson et al., 2002).

More recent, NIR spectroscopy has been applied to detecting defects and authenticating agricultural and food products, such as animal feedstuffs (Murray et al., 2001), pieces of chicken meat (Fumiere et al., 2000), wheat flour (Sirieix and Downey, 1993), Basmati rice (Osborne, Mertens, et al., 1993), coffee beans (Downey and Boussion, 1996), and olive oil (Bertran et al., 2000).

FIGURE 9.3. NIR spectrometer measurement at elevator facilities

An area of growing application for NIR spectroscopy is whole produce quality analysis, which allows traders and consumers to instantly, and nondestructively grade fruits for qualities such as sugar content/sweetness (Bellon-Maurel et al., 1997) or ripeness (Slaughter et al., 1996), and to check for internal defects. Given this already ubiquitous presence, NIR spectroscopy should be a premier choice when confronted by new food analysis tasks such as GMO detection.

To facilitate the reader in understanding why it may be possible for NIR to detect some transgenic foods, a brief explanation of the method, followed by a GMO detection case study, will be presented in the following pages. For more detailed explanations concerning food analysis via NIR spectroscopy, some excellent books and reviews are available (Osborne, Fearn, et al., 1993; Burns and Ciurczak, 2001; Williams and Norris, 2001).

Physics of Near-Infrared Spectroscopy

Definition

Optical spectroscopy is the study of the interaction between electromagnetic radiation and matter. Possible interactions include absorbance, emission, and scattering. The NIR spectrum, which is commonly defined as covering the wavelength range from 700 to 2,500 nm, is positioned between the visible and infrared portions of the electromagnetic spectrum (Figure 9.4). In this region of the elec-

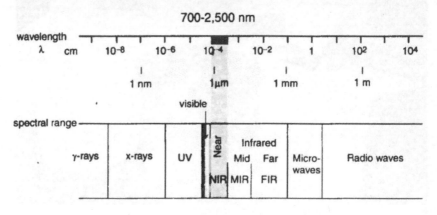

FIGURE 9.4. The near-infrared region (700 to 2,500 nm) in the electromagnetic spectrum

tromagnetic spectrum, the most common interactions are absorbance and scattering.

NIR spectroscopy relies on the relationship between a sample's absorbance of incident NIR radiation, and the concentration of absorbing species within the sample. When NIR radiation impinges on a sample, certain molecules within the sample will vibrate depending on their mass, chemical bonding structure, and the wavelength of the incident radiation. This relationship makes vibrational spectra useful for gaining insight into the molecular structure of a compound. Furthermore, the proportion of the incident light that is absorbed by the sample is proportional to the concentrations of absorbing species within the sample. This relationship is dictated by Beer's law (equation 9.1).

Beer's Law

Beer's law states that the absorption A of light is proportional to the number of absorbing molecules:

$$A = \varepsilon \, l c \qquad (9.1)$$

where c is the concentration of the absorbing molecule, l is the pathlength, and ε is the coefficient of molar absorptivity. Thus, if ε and l are constant, the equation can be rearranged to predict concentration as a linear function of optical absorbance. The coefficient of molar absorptivity for a particular molecule or compound is a function of wavelength and the sample matrix. Where ε is large, the constituent absorbs radiation strongly, and analyses using these wavelengths will be more sensitive. NIR coefficients of molar absorptivity are quite small relative to those for the mid-IR. Although this results in lower sensitivity for NIR, it is advantageous because it allows optical radiation to be transmitted through relatively more thick, intact, and higher moisture concentration samples than would be possible with the mid-IR.

Calibration

For each constituent of any given sample, the shape of the molar absorptivity function is determined by the chemical bonds present in the constituent molecules and their interaction with the sample matrix. For the NIR, the regions of highest molar absorptivity, i.e., absorption bands, are due to the overtones and combinations of the fundamental vibrations of C-H, O-H, and N-H bonds. These absorption bands, along with a myriad of less prominent bands, make up the complete near-infrared spectroscopic profile for a sample, which makes NIR spectroscopy well suited for many organic analysis tasks. However, in the NIR region, significant ambiguity exists in the precise location of absorbance bands, which makes NIR spectra relatively featureless (Figure 9.5). Furthermore, the overtones and combinations of fundamental vibration bands are relatively weak. For these reasons, only water and organic molecules with rather high concentration (~0.1 percent for grain) are generally quantified using NIR spectroscopy.

Because the coefficient of molar absorptivity is rarely known a priori for samples analyzed by NIR, equation 9.1 is usually replaced

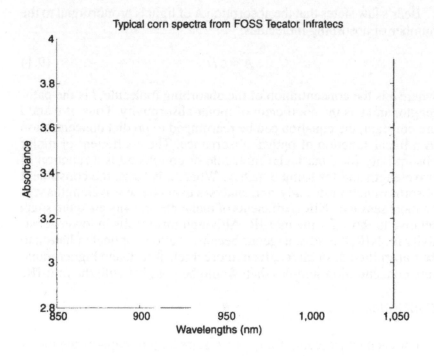

FIGURE 9.5. Typical near-infrared spectra of whole corn kernels

with an empirical calibration equation, obtained via linear regression:

$$c = log(1/T)b_1 + b_0 + e \qquad (9.2)$$

where T is the proportion of incident radiation that is not absorbed by the sample (i.e., the proportion of light that passes through the sample), and b_1, b_0, and e, respectively, are the slope, intercept, and error terms of the calibration equation. Although equation 9.2 assumes measurement by transmitting light through the sample, a similar equation can be derived for reflectance measurements. However, for the sake of simplicity, only transmittance measurements will be discussed herein.

This need for calibration highlights an important characteristic of NIR spectroscopy: it is an *indirect method* that must rely on the quality of another analysis method—a reference method—to be used as a quantitative or qualitative measuring tool. Thus, unlike direct methods of GMO detection, a basis for GMO detection by NIR spectroscopy must be inferred during the calibration process.

For most constituents, however, the limitations of NIR necessitate the measurement of optical absorbance at multiple wavelengths, in addition to the use of multiple linear regression (MLR) to extract enough useful information from a sample spectrum. Much in the same way that a univariate calibration equation (equation 9.2) is derived by solving directly for the least-squares solution, MLR directly solves the least-squares solution to a multiple-wavelength calibration equation, so long as there are more training samples in the calibration set than there are calibration coefficients. Although MLR can generally provide a very accurate, useful solution (Martens and Naes, 1989), the method encounters problems with rotational ambiguity when predictor variables are highly correlated with one another (i.e., multicollinearity), as is the case for the neighboring data points from most full-spectrum NIR analyzers. In these cases, it is necessary to use more complex multivariate calibration techniques.

Instrumentation

Regardless of its intended application, every NIR spectrometer is comprised of (at least) five basic parts: light source, wavelength selector, sample presentation, detector, and signal processor (Figure 9.6).

1. Light source
2. Collimator lens
3. Chopper
4. Wavelength selector: Filter wheel with NIR filters
5. Lens
6. Aperture
7. Tilting mirror
   a. Sample measurement
   b. Reference measurement
8. Integrating sphere
9. Detector and signal processor
10. Optical window
11. Sample

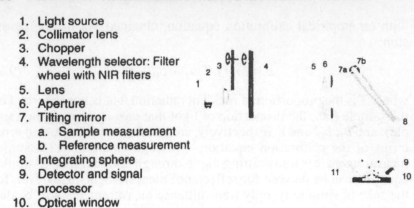

FIGURE 9.6. Schematic example of a commercial NIR analyzer (*Source:* From Bran+Luebbe, Norderstedt, Germany. Web site: <http:www.bran-luebbe.de/deu/tempshort/analyticsweb/en/NIRintro.html>.)

Although the quality and efficiency of most components have improved dramatically in preceding decades, the critical parts of a commercial NIR analyzer look basically the same as they have for the past thirty years. Many of the improvements in NIR spectroscopy have actually been a product of the digital revolution, rather than developments in instrumentation. Advances in digital storage and signal processing have allowed the creation of vast calibration databases, and the utilization of complex mathematical algorithms, both of which have been instrumental in promoting NIR in recent years. Furthermore, because of recent growth in the digital video and telecommunications industries, which require many of the components that are used for NIR spectroscopy, the cost of NIR instrumentation is relatively low.

Chemometrics

The multivariate modeling requirements of NIR spectroscopy have helped spawn a new scientific field called chemometrics. *Chemometrics* is the chemical discipline that uses mathematics and statistics to design or select optimal experimental procedures, provide maximum relevant chemical information by analyzing chemical data, and obtain knowledge about chemical systems (Massart et al., 1998). Those interested in chemometrics should consult Martens and Naes, 1989; Massart et al., 1998; Mobley et al., 1996; Workman et al., 1996; and Bro et al., 1997, which are acclaimed reviews of routine and advanced chemometrics concepts.

Of the many chemometric modeling techniques available, partial least squares (PLS) regression (Geladi and Kowalski, 1986) is the one most often used for creating analytical models from NIR spectra. This widespread popularity is due to its combination of analytical power, ease of use, and good interpretability. PLS is similar to MLR in that a linear calibration equation is created. However, many differences exist in the method each algorithm uses to arrive at a solution.

Whereas MLR solves directly for the least-squares fit using all the spectral data given, PLS first reduces the data by projecting the spectra into a subset of orthogonal basis functions, generally called factors, much in the same way as principal components analysis (PCA) (Cowe and MacNicol, 1985). However, PLS factors are automatically arranged in order of decreasing correlation with the predicted variable, which eases the decision of which factors to choose. Once a suitable number of factors have been selected, and the data have been projected into the factor subspace, a calibration equation is created by regressing the factor scores against the constituent to be predicted. Practitioners find it advantageous to use PLS regression because it lessens the need to determine the best wavelengths for modeling, and can be used when there are fewer training data points than model coefficients, and only one parameter to optimize (number of factors).

In addition to PLS, several alternative linear modeling strategies have been devised: principal component regression (Naes and Martens, 1988) and robust regression (Rousseeuw and Leroy, 1987), along with some nonlinear modeling strategies, such as locally weighted regression (LWR) (Naes and Isaksson, 1992) and artificial neural networks (ANN) (Fiesler and Beale, 1997). Many of these modeling strategies will be more or less applicable depending on the experience of the practitioner and the complexity of the problem. Indeed, even some of the simplest chemometric modeling packages can be used to develop a "black-box" model of nearly any data for which a correlation can be found. For example, chemometric regression techniques can be used for pattern recognition issues, such as discriminate analysis, that have no quantitative reference values. Instead, a classification model can be developed to discriminate between two groups of samples using a qualitative reference, such as GMO status.

The preceding paragraphs have illustrated some of the benefits of NIR spectroscopy for food analysis: quantitative and qualitative analytical power, nondestructive nature, little or no sample preparation, rapidity, ease of use, and low cost. Despite relative limitations in sensitivity, and being an indirect method, the potential advantages of using NIR spectroscopy for detecting genetically modified organisms in grain justify an attempt for its application to GMO detection. The following case study used NIR spectroscopy to develop models that were able to correctly classify whole grain samples of Roundup Ready and non–Roundup Ready soybeans (Roussel et al., 2001).

ROUNDUP READY SOYBEAN DETECTION CASE STUDY

Background

In response to the growing public and regulatory concerns in the United States and abroad, the Grain Quality Laboratory (http://www. iowagrain.org) of Iowa State University began a feasibility study in 1999 to investigate possible strategies for detecting Roundup Ready (RR) soybeans in whole grain samples using NIR spectroscopy. At that time, the RR portion of total soybean acreage was growing rapidly (from 0 percent of soybean acres in 1995, to more than 50 percent in 2000) (NIST, 2000), and the grain handling industry had yet to develop a credible, comprehensive GMO screening strategy based on direct GMO detection methods.

If enough consistent spectral difference existed between RR and non–Roundup Ready (NRR) soybeans, NIR would be well suited to screening inbound grain deliveries. Although NIR spectrometers are not precise enough to detect compounds at the DNA concentration level (parts per trillion), spectral differences caused by larger structural changes (if any) accompanying the modification might be measurable.

The prospects of GMO detection via NIR spectroscopy were motivated by the results of a preliminary test conducted by the GQL in 1998, which suggested a spectral difference between RR and NRR soybeans. A calibration, derived using the 1998 harvest data, classified RR soybeans with 85 percent accuracy, and NRR soybeans with 95 percent accuracy when tested on an independent sample set. However, when the 1998 model was applied to data from the next harvest year, classification accuracy was greatly reduced. More samples and growing seasons were needed to develop a reliable RR soybean screening method using NIR spectroscopy.

Hence, the objective of this feasibility study was to develop a protocol for distinguishing Roundup Ready soybeans from non–Roundup Ready soybeans using NIR spectroscopy. The work compared three chemometric discriminate analysis techniques using data from two soybean-growing seasons.

Materials

NIR spectroscopy is commonly used to perform rapid, nondestructive whole grain proximate analysis. FOSS Tecator Infratec whole grain analyzers are the NIR spectrometers approved for official USDA analysis of various grains, including wheat, corn, and soybeans (Figure 9.7) (AACC, 1999; NIST, 2000). It is also the instru-

FIGURE 9.7 Officially approved NIR spectrometer for grain analyses in the United States: FOSS Tecator Infratec 1229 Grain Analyser

ment model used most commonly at grain buying stations that handle soybeans. For grain handlers to screen for RR soybeans using technology they already own would present a significant cost and time savings over their adoption of direct methods of RR soybean detection.

The spectral data for this study were collected using three Infratec whole grain analyzers (2 Infratecs, model 1229: SN #553075 and SN #553337; 1 Infratec Model 1221: SN #230966). The analyzers collected NIR spectral data by measuring the relative amount of light passed through a 30 mm column of whole grain (soybeans). The spectra were acquired via scanning monochrometer in 2 nm increments, from 850 to 1,048 nm (100 spectral data points), with a 10 nm bandpass (Figure 9.8).

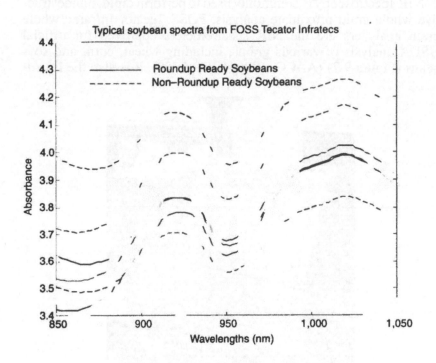

FIGURE 9.8. Example of near-infrared spectra of Roundup Ready and non–Roundup Ready soybeans

A spectral database was created, consisting of spectral scans from 8,829 samples. Small portions of the samples were collected from various independent and seed-breeder plots located around Iowa during the 1998 and early 1999 harvest. These samples were scanned using Infratec SN #553075. The majority of the sample database originated from Iowa State University Yield Trial's 1999 harvest. Every variety represented in the trial was planted at each of nine locations spread evenly across Iowa. The spectra from the ISU yield test were acquired using Infratecs SN #553337 and #230966.

The data were balanced in relation to major constituents known to affect the spectra (moisture, protein, oil, and fiber), between classes, and between instruments (Table 9.1). This was done to block any secondary correlations between soybean composition and GMO status that might have influenced classification performance, that is, to avoid basing the classification model on an artifact, such as a moisture difference between RR and NRR. In all cases, the variety information was provided by the sample originator, and was used as the reference (true) RR/NRR status for discriminate modeling. Some ex-

TABLE 9.1. Composition of sample sets

NIR unit	Samples (RR/NRR)	Constituent[a]	Roundup Ready		Conventional soybeans	
			Average	SD[b]	Average	SD[b]
Spectrometer #1	breeder,	moisture	11.2	0.63	10.7	1.71
(# 553075)	strip plots	protein	36.0	1.50	35.8	1.61
	(308 / 341)	oil	18.5	0.82	18.3	1.00
Spectrometer #2	Iowa soybean	moisture	8.8	1.13	9.3	1.18
(# 230966)	yield test	protein	35.1	1.03	34.8	1.08
	(2,562 / 2,147)	oil	18.7	0.64	18.7	0.61
Spectrometer #3	Iowa soybean	moisture	8.7	0.97	9.1	0.83
(# 553337)	yield test	protein	36.2	1.21	36.1	1.48
	(1,738 / 1,733)	oil	17.8	0.80	18.0	0.80
All Spectrometers	All	moisture	8.9	1.18	9.3	1.10
	(4,622 / 4,238)	protein	35.8	1.34	35.6	1.52
		oil	18.2	0.87	18.3	0.83

[a]All constituents were predicted by the FOSS Tecator Infratec, calibration SB009904, on a 13 percent moisture basis.
[b]SD: standard deviation.

cessively dry samples originating from seed-breeder plots had to be rehydrated prior to NIR analysis.

Methods

To fulfill the objectives of this study, classification models were created using three discriminate analysis strategies: partial least squares regression (PLS), locally weighted regression, and artificial neural networks. For each classification model, calibrations were derived using a randomly selected calibration set, and tested on the remaining validation samples. Model performance was compared by relating percent *correct* classification RR and NRR.

Partial Least Squares

Partial least squares discriminate analysis (PLS-DA) models (Sjöström et al., 1986) were derived using Unscrambler, version 7.5 (CAMO, Inc.), whereas all other chemometric algorithms were implemented in MATLAB, version 5.3 (The MathWorks, Inc.). The number of PLS factors to retain for each model was tuned using cross-validation. All chemometric models used a conjunctive Boolean coding scheme (i.e., one column for both classes) where all Roundup Ready (RR) samples were assigned a "1" and non–Roundup Ready (NRR) samples were assigned a "0." Since the class sizes were well balanced in every sample subset, 0.5 was used as the class threshold for predicted values.

Although PLS regression attempts to derive the optimal linear calibration equation to fit the entire space spanned by the calibration database, it is likely that different calibration equations would have a much better fit in some subspaces of the calibration data. Thus, it would be beneficial to either derive a more complex, nonlinear calibration equation that can fit the various localities of the calibration space (for instance, using an artificial neural network model), or derive many local linear calibration equations, each being particular to a location in calibration space (local-linear regression).

Locally Weighted Regression

Locally weighted regression is a nonlinear prediction technique based on the concepts of local-linear regression and weighted principal components regression (Naes and Isaksson, 1992; Wang et al., 1994). Instead of storing a calibration equation (e.g., as is done for linear calibration), the calibration database is compressed using PCA and retained in memory. For prediction, a unique calibration equation is derived for each sample, with only the nearest calibration data points being used.

Thus, in order to implement LWR, it was necessary to manually tune the number of principal components (factors), and the number of calibration samples that would be used for local regression subsets (number of nearest neighbors to select).

Artificial Neural Networks

Artificial Neural Network modeling has become a popular method of deriving global, nonlinear regression equations, and is the main alternative to local modeling. Although its advantage (or disadvantage) versus local modeling is debatable, in terms of performance, it is an easier method to implement using the analyzers that are available currently, since only a single set of coefficients must be stored, instead of the entire calibration database.

An ANN regression equation consists of a collection of nodes (or neurons) connected by a series of weight coefficients (Figure 9.9).

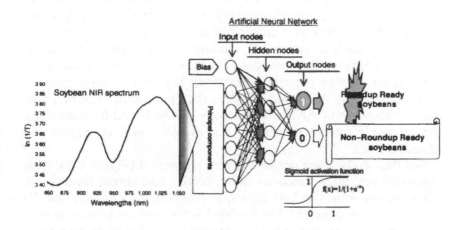

Figure 9.9. Artificial neural network architecture

ANN predictions progress, layer by layer, from the input nodes, through the layer(s) of hidden nodes, to the output nodes. At the input layer, the nodes can be raw data points or some other preprocessed form of the data. For NIR spectroscopy, the inputs are often a subset of spectral principal components scores in order to avoid problems of multicollinearity. At each subsequent layer, the node values are a function of the sum of the connected nodes in the previous layer multiplied by their respective weight coefficient. The node activation function can be linear or nonlinear, and must be selected prior to derivation of the ANN model (Figure 9.9).

An ANN is trained by tuning its weight coefficients according to squared error of prediction until some stopping criterion is reached. Unlike linear least squares regressions, such as PLS, which arrive at a unique, explicit solution by making certain assumptions about the distribution of calibration data, ANN interactively derives an implicit solution using error-gradient backpropagation (Rumelhart, Hinton, and Williams, 1986), making no assumptions about the distribution of the calibration data. Because the ANN derivation process can be prone to generating only locally minimal solutions, it is necessary to repeatedly train and test the model using random starting weights in an attempt to create the best possible model.

It is clear that model complexity grows quickly as nodes are added. Thus, different preprocessing techniques and training algorithms are used to avoid deriving an excessively complex model from the training data (due to the high number of trained weights, an ANN could "memorize" the calibration set), leading to poor performance when making predictions from new data. Because of the vast amount of peripheral knowledge required to competently derive ANN models, they are generally developed only by expert practitioners. A more in-depth discussion of ANN methodology can be found by consulting the following references: Fiesler and Beale, 1997; Despagne and Massart, 1998.

For this study, three network layers were used (Figure 9.9), with sigmoidal activation functions in the hidden and output layers. The ANN inputs were the first principal components (20, 25, and 30 components were tested). The reported ANN performance is the *best* achieved during a series of random training runs. Based on the results of previous work (Roussel et al., 2000), the ANN was trained using a modification of the error-gradient backpropagation strategy called *regularization learning* (Girosi et al., 1995). Regularization learning adds a second term to the error function that is the sum of all network weights. In this way, the solution with the lowest error and the lowest sum of weights is found during network training. Because of this effect, regularization learning produces "smoother" networks that are often more able to generalize results.

Results

Table 9.2 shows a summary of classification results according to modeling technique. PLS was able to correctly classify 75 percent of the samples for each class. Although these results indicate some spectral feature that can be modeled, such performance is hardly useful.

The direction of some spectral differences between Roundup Ready (RR) and non–Roundup Ready (NRR) soybeans were consistently positive (RR higher absorbance than NRR), but the magnitude of the difference was not consistent among sample origination groups. Soybean composition has long been known to be environmentally variable. The component responsible for an effect that can be detected via NIR spectroscopy may also vary according to growing environment. However, the nonlinearity added by an inconsistent effect may preclude the use of linear discriminate analysis techniques, such as PLS.

TABLE 9.2. Summary of classification accuracy for Roundup Ready soybean discrimination models

Calibration set: Number of samples (RR/NRR)	Predicted test set: Number of samples (RR/NRR)	Model	Parameters	% correct classification: Global	RR	NRR
7,919 samples	910 samples	PLS	16 factors	75	75	74
		LWR	18 factors 700 neighbors	91	89	93
(4,151/3,768)	(457/453)	ANN	20:8:2[a]	89	89	88
		LWR+m	18 factors 400 neighbors	96	94	97

[a]20 input notes (principal components); 8 hidden nodes; 2 output nodes

This nonlinearity seems to be suggested by the considerably better performance achieved by LWR and ANN, both of which are suited to modeling nonlinear relationships. Although LWR was slightly better at identifying NRR samples (93 percent and 89 percent of correct classification for NRR and RR, respectively), the results for LWR and ANN were essentially the same, with averages of 91 percent and 89 percent correct classification, respectively. In the work conducted at the Grain Quality Laboratory leading up to this study, LWR generally outperformed ANN. Moreover, the gentle shape of the LWR optimization response surface (Figure 9.10) shows some insensitivity to changes in parameter settings, which suggests some degree of model robustness. Furthermore, in later tests using the same database and a modified version of the LWR algorithm (which selected local calibration samples based on moisture as well as Mahalanobis distance) average predictive performance was increased to 96 percent.

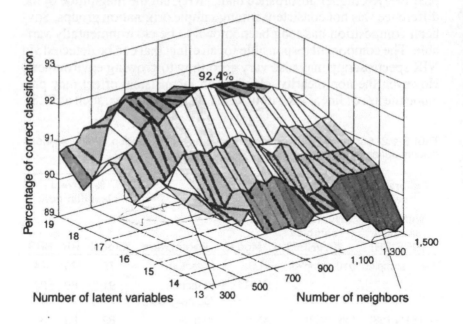

FIGURE 9.10. Influence of locally weighted regression parameters on Roundup Ready soybean classification. (*Source:* Reprinted from *Applied Spectroscopy,* Vol. 55, Roussel et al., Detection of Roundup Ready soybeans by near-infrared spectroscopy, pp. 1425-1430, Copyright 2001, with permission from the Society for Applied Spectroscopy.)

Conclusions and Discussion

A number of conclusions were drawn from this feasibility study and associated works:

1. Samples of Roundup Ready soybeans can be distinguished from non–Roundup Ready samples using NIR spectroscopy and either locally weighted regression or artificial neural network modeling.
2. Locally weighted regression tends to perform slightly better than artificial neural networks and partial least squares.
3. The source of the Roundup Ready spectral effect is unknown, but the direction of the spectral difference is consistent across crop years.
4. In situations where NIR spectroscopy is already being used for soybean quality analysis, it is likely that an added GMO screen (via NIR grain analyzer) could be conducted with little or no additional financial or time cost.

However, many technical and practical questions remain that must be answered by further research studies concerning the use of NIR spectroscopy to detect Roundup Ready soy in food. For example:

1. If NIR spectroscopy is used for Roundup Ready soybean detection, what new regulatory challenges will be created (i.e., who will create and maintain calibration models, what instruments will be legal, etc.)?
2. Would further preprocessing of the spectral data or the use of some other chemometric method improve on the results achieved in this study?
3. Can NIR detect the presence of Roundup Ready soy in processed food?

QUALITY ASSURANCE ISSUES

As is the case for any new analytical method, some work should be done to assure the quality of (i.e., "validate") GMO screens that rely on NIR spectroscopy. Since NIR spectroscopy is an indirect technique, method validation can be a complex task. A comprehensive

NIR validation package should include both model-specific and instrumental details.

Model Validation

Model validation is a process that is fairly unique to indirect analytical techniques, which require calibration. Validating an NIR spectroscopic model entails rigorous testing of the calibration model performance and interpretation of the model parameters.

1. *Model testing:* to have confidence in future predictions made using an NIR model, the model should be thoroughly tested for (among others) accuracy, limit of detection, specificity, and robustness.

   a. *Accuracy* is generally defined as the root mean squared error of prediction (standard error of prediction: SEP), computed on an independent (different batch, etc.) test set or by cross-validation (Efron, 1982). Method accuracy is the summation of expected errors due to the precision of the instrument, precision of the reference data, and the model's lack of fit (Osborne et al., 1993).

   b. *Limit of detection (LOD)* is the smallest unit of an analyte that can be detected repeatedly, and is related to the precision of the instrument. Although not often cited as a measure of model performance, it is important that LOD be established to avoid undue confidence in the model, in particular for GMO detection.

   c. *Specificity* has been defined as "the ability to assess unequivocally the analyte in the presence of components which may be expected to be present" (FDA, 2000). Although an interpretation of this definition would suggest that extensive and diverse testing could validate the specificity of an NIR GMO detection model, it could present a challenge for NIR spectroscopy if taken literally. In the case of detecting Roundup Ready soy, for example, one problem is that no "analyte" has been defined except the quality of being Roundup Ready.

   d. *Robustness* is the stability of the model accuracy under a change in experimental conditions (Vander Heyden et al., 2001), such as different instruments or operators, or when analyzing "extreme samples." For a GMO screen using NIR spectroscopy, the model must be robust enough to be implemented in many uncontrolled locations. Furthermore, it may be necessary to determine the robustness of the model to changes in the sample phenotype, variety, or growing season, so that its useful limits can be set.

2. *Model interpretation:* any model created using NIR spectroscopy will have many parameters that may require interpretation depending on the intended use. These parameters include descriptive statistics concerning the calibration database and model equation coefficients.

   a. *Database statistics* define the useful range and the variability that has been built into the calibration model. The size of the database and the number of outliers by themselves can give some indication of model generality. It is generally held that as more data are added to an NIR calibration database, model performance will improve until it is limited by the quality of the reference chemistry and instrument precision.

   b. *Model coefficients* and training parameters (i.e., number of factors included, etc.) can be used to gain insight into the source of model specificity. The larger regression coefficients of a spectroscopic calibration model should correspond to the wavelength regions of high molar absorptivity for the analyte of interest. This information can help confer a basis of chemical understanding, or legitimacy, upon the calibration. This is difficult for qualitative and discriminate analysis calibrations, such as GMO detection, that may not have a clear "analyte." Thus, although large model coefficients may allude to some chemical change, the ambiguity in NIR spectra does not always allow unknown absorbance bands to be identified with certainty (Figure 9.11).

Instrument Validation

Once an NIR model has been thoroughly evaluated and (hopefully) understood, it is important to consider a number of instrumental effects that will influence model performance, including (among others) instrumental stability, standardization, and operational considerations.

1. *Instrument standardization:* even more so than for other analytical methods, instrument standardization is an important problem for NIR spectroscopy because of its reliance on complex, empirical calibrations. In the case of GMO detection in grain,

networks of hundreds of NIR grain analyzers must be standardized. However, the need for standardization can sometimes be
reduced using spectral preprocessing techniques, or by including samples in the model from many instruments to increase the
generalization power of the model (Fearn, 2001).

2. *Instrument stability* is related to the problem of instrument standardization. Some NIR analyzers may be affected by changes in
environment, or they may be prone to drift as optical components age. To ensure that maximum calibration performance is
achieved, and to have confidence in predictions, the stability of
an instrument must be assessed. The necessary level of stability,
however, is dependent on the complexity of the application.

3. *Operational considerations:* Even when an NIR analyzer has
been correctly standardized and is stable, in some instances
method performance can be influenced significantly in the way
the instrument is operated. However, no such influence has been
observed thus far for NIR grain analyzers. Thus, unlike the direct method, no skilled operator should be required for an NIR
solution.

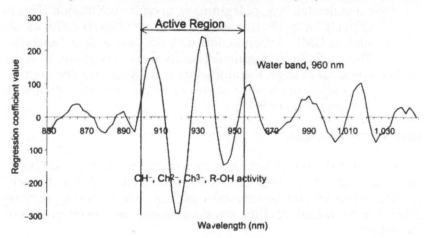

Regression coefficients for 1998 Roundup Ready soybean calibrations

FIGURE 9.11. Regression coefficients of Roundup Ready soybean classification
models. (*Source:* Reprinted from *Applied Spectroscopy,* Vol. 55, Roussel et al.,
Detection of Roundup Ready soybeans by near-infrared spectroscopy, pp. 1425-
1430, Copyright 2001, with permission from the Society for Applied Spectroscopy.)

Practical Considerations for GMO Detection via NIR

Since it was intended to be used only as a feasibility study that would guide future work, the case study described in this chapter utilized only a few of the methods listed previously for validating predictions. A large, diverse calibration database was used, which included multiple instruments and samples from two growing seasons. Lacking from the study, however, was a completely independent test set (i.e., a test set coming from a third growing season) to assess model performance more confidently. Moreover, it is well known that, for the grain applications, NIR spectra from several years must be included in the models to take into account the year-to-year variability (growing conditions, events, etc.). Furthermore, no work was done to establish limits on detection (particularly for blended grain samples).

As mentioned previously, assessing the specificity of a GMO detection model using NIR spectroscopy is a paradoxical problem. To literally define the specificity of the method, an actual "analyte," that is, an identified molecular dissimilarity between a GM and non-GM organism, must be known. The existence of this "analyte," if concentrated enough to be detectable by NIR spectroscopy, however, defies the notion of "substantial equivalency" (Maryanski, 1995), which states that the nutritional and toxin content of those GM products should fall within the range of concentrations normally observed in conventional varieties. However, since it must be assumed that an NIR food analyzer is incapable of measuring DNA changes, with some exceptions, it should only be possible to detect a GMO (using NIR) if it violates this notion. Exceptions to this would be if NIR GMO detection was relying on phenotypic differences that were correlated to GM status or if the limit of detection of NIR spectroscopy is better than has usually been assumed. The resolution of this question, however, will likely be answered only with the help of direct chemical assays and more experimentation that will either debunk or validate the NIR model of GMO detection.

FUTURE CONSIDERATIONS

NIR spectroscopy could significantly reduce the burden of implementing direct methods (e.g., DNA-based or protein-based screens) of GMO detection by augmenting their capabilities as a *prescreen*. For example, a calibration might be derived to identify samples that have a very high probability of being non-genetically modified, leaving all other samples (with claims of purity) to be screened using direct methods, which are more suitable to resolve cases of legal dispute. This way, it would be possible to reduce the cost and time burden of GMO detection while minimizing losses due to error.

The question remains as to what other GMO events can be detected using NIR spectroscopy. It would be difficult to answer that question for all GMO events, but some preliminary work done at the Grain Quality Laboratory gives hope to the idea of discriminating Bt corn using NIR spectroscopy.

In particular, a study was performed to create a calibration for discriminating between *individual kernels* of Bt and non-Bt corn. For the study, 359 kernels (198 Bt, 161 non-Bt) were analyzed using an NIR hyperspectral imaging spectrometer (Cogdill et al., 2003; Cogdill et al., 2002). The kernels used for the study were of diverse origin, both within and between the two classes of samples, year of harvest, number of hybrids represented, and kernel morphology. The spectrometer was able to classify kernels with an average accuracy (in cross-validation) of nearly 99 percent, using PLS. However, because the imaging spectrometer remains in development, no further tests were conducted. Single-seed NIR grain analysis would be useful for handling grain with mixed-in GMOs because it can directly estimate the proportion of genetically modified seeds within a sample.

Another exciting prospect for GMO screening using NIR spectroscopy is online detection. For example, a network of NIR analyzers could be installed at strategic points throughout a grain elevator, providing constant monitoring of the grain operations. This may someday give elevator operators another chance to avoid GMO contamination as a result of routing errors.

Although these potential developments could add to the productivity of handling segregated grain, neither would likely be possible using direct methods of GMO detection.

Many more hurdles of technology and understanding must be cleared before NIR spectroscopy will be widely accepted as a means of detecting GMOs. However, as advances in instrument technology and chemometric techniques continue, NIR spectroscopy will become an even more attractive option for food analysis tasks, and possibly for detecting genetically modified organisms in food and food ingredients.

REFERENCES

American Association of Cereal Chemists (AACC) (1999). *Method 39-01: General near-infrared instrumentation and techniques; Method 39-21: Near-infrared method for whole-grain analysis.* AACC, St. Paul.

Barton, F.E. II (2002). Theory and principles of near infrared spectroscopy. *Spectrosc. Europe* 14(1):12-18.

Bellon-Maurel, V., Steinmetz, V., Dusserre-Bresson, L., Jacques, J.C., and Rannou, G. (1997). Real-time NIR sensors to sort fruit and vegetables according to their sugar content. In *International Symposium on Fruit Nut and Vegetable Production Engineering*, Davis, California.

Berntsson, O., Danielsson, L.G., Lagerholm, B., and Folestad, S. (2002). Quantitative in-line monitoring of powder blending by near infrared reflection spectroscopy. *Powder Technol.* 123(2-3):185-193.

Bertran, E., Blanco, M., Coello, J., Iturriaga, H., Maspoch, S., and Montoliu, I. (2000). Near infrared spectrometry and pattern recognition as screening methods for the authentication of virgin olive oils of very close geographical origins. *J. Near Infrared Spectros.* 8:45-52.

Bro, R., Workman, J.J., Mobley, P.R., and Kowalski, B.R. (1997). Review of chemometrics applied to spectroscopy: 1985-1995, part III: Multi-way analysis. *Appl. Spectrosc. Rev.* 32:237-261.

Burns, D.A. and Ciurczak, E.W., eds. (2001). *Handbook of Near-Infrared Analysis*, Second Edition, Revised and Expanded, *Practical Spectroscopy Series*, Volume 27. Marcel Dekker Inc., New York.

Cogdill, R., Hurburgh, C., Jensen, T.C., and Jones, R.W. (2002). Single kernel maize analysis by near-infrared hyperspectral imaging. In *Near Infrared Spectroscopy: Proceedings of the 11th International Conference*. NIR Publications, Chichester, UK (in press).

Cogdill, R., Hurburgh, C., and Rippke, G.R. (2003, submitted). Single kernel maize analysis by near-infrared hyperspectral imaging. *Trans. ASAE.*

Cowe, I.A. and MacNicol, J.W. (1985). The use of principal components in the analysis of near-infrared spectra. *Appl. Spec.* 39(2):257-266.

Despagne, F. and Massart, D.L. (1998). Tutorial review: Neural networks in multivariate calibration. *The Analyst* 123:157R-178R.

Downey, G. and Boussion, J. (1996). Authentication of coffee bean variety by near-infrared reflectance spectroscopy of dried extract. *J. Sci. Food Ag.* 71(1):41-49.

Efron, B. (1982). *The Jackknife, the bootstrap, and other resampling plans.* Society for Industrial and Applied Mathematics, Philadelphia, Pennsylvania.

Fearn, T. (2001). Standardisation and calibration transfer for near infrared instruments: A review. *J. Near Infrared Spectros.* 9:229-244.

Fiesler, E. and Beale, R. (1997). *Handbook of Neural Computation.* IOP Publishing Ltd., and Oxford University Press, Inc., Bristol and Oxford, United Kingdom.

Food and Drug Administration (FDA) (2000). Analytical procedures and method validation (draft guidance). *Fed. Reg.* 65(169):776-777.

Fumiere, O., Sinnaeve, G., and Dardenne, P. (2000). Attempted authentication of cut pieces of chicken meat from certified production using near infrared spectroscopy. *J. Near Infrared Spectros.* 8(1):27-34.

Geladi, P. and Kowalski, B.R. (1986). Partial least squares regression: A tutorial. *Anal. Chem. Acta* 185:1-17.

Girosi, F., Jones, M., and Poggio, T. (1995). Regularization theory and neural networks architectures. *Neural Comp.* 7(2):219-269.

Herschel, W. (1800). Experiments on the solar, and on the terrestrial rays that occasion heat, Part I and Part II. *Phil. Trans. Roy. Soc. (London)* 90:293-326, 437-538.

Hurburgh, C.R. Jr. (1999a). The GMO controversy and grain handling for 2000. In *ISU Integrated Crop Management Conference,* Ames, Iowa.

Hurburgh, C.R. Jr. (1999b). Supply management networks for grains. In *11th Annual Food Focus, Association of Cereal Chemists,* N.W. Section, St. Paul, Minnesota.

Martens, H. and Naes, T. (1989). *Multivariate Calibration.* Wiley, Chichester, United Kingdom.

Maryanski, J.H. (1995). Food and Drug Administration policy for foods developed by biotechnology. In *Genetically Modified Foods: Safety Issues* (pp. 12-22). Engel K.-H., G.R. Takeoka, and R. Teranishi, eds. American Chemical Society, Washington, DC.

Massart, D.L., Vandeginste, B.G.M., Buydens, L.M.C., Jong, S.D., Lewi, P.J., and Smeyers-Verbeke, J. (1998). *Handbooks of Chemometrics and Qualimetrics* (Parts A and B). Elsevier, New York.

Millmier, A., Lorimor, J., Hurburgh, C., Hattey, J., and Zhang, H. (2000). Near-infrared sensing of manure nutrients. *Trans. American Society Electrical Engineers* 43:903-908.

Mobley, P.R., Workman, J.J., Kowalski, B.R., and Bro R. (1996). Review of chemometrics applied to spectroscopy: 1985-1995, part II. *Appl. Spectrosc. Rev.* 31(4):347-368.

Murray, I., Aucott, L.S., and Pike, I.H. (2001). Use of discriminant analysis on visible and near infrared reflectance spectra to detect adulteration of fishmeal with meat and bone meal. *J. Near Infrared Spectros.* 9(4):297-311.

Naes, T. and Hildrum, K.I. (1997). Comparison of multivariate calibration and discriminant analysis in evaluating NIR spectroscopy for determination of meat tenderness. *Appl. Spec.* 51(3):350-357.

Naes, T. and Isaksson, T. (1992). Locally weighted regression in diffuse near infrared transmittance spectroscopy. *Appl. Spec.* 46(1):34-43.

Naes, T. and Martens, H. (1988). Principal component regression in NIR analysis: Viewpoints, background details, and selection of components. *J. Chemom.* 2: 155-167.

National Institute of Standards and Technology (NIST) (2000). *Handbook 44.* U.S. Government Printing Office, Washington, DC.

Norris, K.H. (1964). Design and development of a new moisture meter. *Trans. American Society Electrical Engineers* 45(7):370-372.

Osborne, B.G., Fearn, T., and Hindle, P.H. (1993). *Practical NIR Spectroscopy with Applications in Food and Beverage Analysis,* Second Edition. Longman Scientific and Technical, Harlow, Essex, United Kingdom.

Osborne, B.G., Mertens, B., Thompson, M., and Fearn, T. (1993). The authentication of Basmati rice using near infrared spectroscopy. *J. Near Infrared Spectros.* 1:77-83.

Reeves, J.B. (2001). Near-infrared diffuse reflectance spectroscopy for the analysis of poultry manures. *J. Ag. Food Chem.* 49(5):2193-2197.

Rodriguez-Saona, L.E., Fry, F.S., McLaughlin, M.A., and Calvey, E.M. (2001). Rapid analysis of sugars in fruit juices by FT-NIR spectroscopy. *Carbohyd. Res.* 336(1):63-74.

Rousseeuw, P. and Leroy, A. (1987). *Robust Regression and Outlier Detection.* Wiley, New York.

Roussel, S.A., Hardy, C.L., Hurburgh, C.R. Jr., and Rippke, G.R. (2001). Detection of Roundup Ready soybeans by near-infrared spectroscopy. *Appl. Spec.* 55(10): 1425-1430.

Roussel, S.A., Rippke, G.R. and Hurburgh, C.R. Jr. (2000). Detection of Roundup Ready soybeans by near-infrared spectroscopy. In *10th International Diffuse Reflectance Conference (IDRC2000),* F.E. Barton, ed. Chambersburg, Pennsylvania.

Rumelhart, D.E., Hinton, G.E., and Williams, R.J. (1986). Learning internal representations by error propagation. In *Parallel Distributed Processing* (pp. 318-362) D. Rumelhart, ed. MIT Press, Cambridge, United Kingdom.

Sasic, S. and Ozaki, Y. (2000). Band assignment of near-infrared spectra of milk by use of partial least-squares regression. *Appl. Spec.* 54(9):1327-1338.

Shenk, J.S. and Westerhaus, M.O. (1993). Near-infrared reflectance analysis with single-product and multiproduct calibration. *Crop Sci.* 33(3):582-584.

Sinnaeve, G., Dardenne, P., Agneessens, R., and Biston, R. (1994). The use of near infrared spectroscopy for the analysis of fresh grass silage. *J. Near Infrared Spectros.* 2:79-84.

Sirieix, A. and Downey, G. (1993). Commercial wheatflour authentication by discriminant analysis of near infrared reflectance spectra. *J. Near Infrared Spectros.* 1:187-197.

Sjöström, M., Wold, S., and Söderström, B. (1986). PLS discriminant plots. In *Pattern Recognition in Practice II* (pp. 461-470), E.S. Gelsema and L.N. Kanal, eds. Elsevier, Amsterdam, the Netherlands.

Slaughter, D.C., Barrett, D., and Boersig, M. (1996). Nondestructive determination of soluble solids in tomatoes using near infrared spectroscopy. *J. Food Sci.* 61(4):695-697.

United States Department of Agriculture (USDA) (1995). *Grain Inspection Handbook.* Grain Inspection, Packers and Stockyards Adminisatration (GIPSA), Federal Grain Inspection Services (FGIS), Washington, DC.

Vander Heyden, Y., Nijhuis, A., Smeyers-Verbeke, J., Vandeginste, B.G.M., and Massart, D.L. (2001). Guidance for robustness/ruggedness tests in method validation. *J. Pharm. Biomed. Anal.* 24(5-6):723-753.

Wang, Z., Isaksson, T., and Kowalski, B.R. (1994). New approach for distance measurement in locally weighted regression. *Anal. Chem.* 66:249-260.

Williams, P. and Norris, K., eds. (2001). *Near-Infrared Technology in the Agricultural and Food Industries,* Second Edition. AACC, St. Paul.

Williams, P.C. and Sobering, D.C. (1993). Comparison of commercial near infrared transmittance and reflectance instruments for analysis of whole grains and seeds. *J. Near Infrared Spectros.* 1:25-32.

Workman, J.J., Mobley, P.R., Kowalski, B.R., and Bro, R. (1996). Review of chemometrics applied to spectroscopy: 1985-95, part I. *Appl. Spectrosc. Rev.* 31(1-2):73-124.

Chapter 10

Other Methods for GMO Detection and Overall Assessment of the Risks

Farid E. Ahmed

This chapter details detection methods not mentioned in earlier chapters, expands on discussion of important issues related to gene flow and intellectual property rights, discusses examples of the different safety regimens employed in the United States and the European Union (EU), and concludes by giving a critique of the methods commonly used to test for genetically modified organisms (GMOs).

METABOLIC PROFILING

GM crops can be identified by alterations in the cellular content of a wide variety of plant metabolites, approaching 200,000 compounds. This holistic approach to analysis has been enhanced by advances in gas chromatography (GC), high-performance liquid chromatography (HPLC), mass spectrometry (MS), nuclear magnetic resonance (NMR), and Fourier-transformation near-infrared spectroscopy (FTNIRS) (Hall et al., 2002). Applications can be either unbiased or targeted toward metabolites in certain metabolic pathways or compound classes (Kuiper et al., 2003).

GC-MS combinations have been widely used because GC results in highly resolved chromatograms, and MS provides information on internal structures of analyzed compounds (Roessner et al., 2001). However, the identification of unknown metabolites is difficult when employing a GC analysis because the high temperatures needed to

---

I express my gratitude to colleagues who permitted me to use their figures, which enhanced the readability of this chapter.

volatilize the compounds may cause degradation of unstable metabolites. This difficulty has resulted in the use of HPLC to overcome this limitation (Tolstikov and Fiehn, 2002). Another technique involves NMR for the analysis of crude GMO extracts. Proton NMR and FTNIRS have also been applied to metabolic fingerprinting of GMOs (Charlton et al., 2004; Kuiper et al., 2003) without the need for compound identification.

The nontargeted approaches for metabolic profiling represent a compromise between comprehensibility and specificity of detection. Limitations such as nonavailability of reference material and the need for appropriate bioinformatics tools for data analysis represent challenges to the wide application of these techniques for GMO detection. These methods, however, are well suited to the detection of second and third generation transgenics that have altered profiles for nutritional components.

PROTEOMICS

Proteome is the protein complement of the genome. *Proteomics* is the large-scale study of protein properties (e.g., expression level, posttranslational modification, interactions) to gain an understanding of gene function at the translational level (Blackstock and Weir, 1999). Unlike genomics, proteomics has no equivalent of amplification by polymerase chain reaction (PCR), thus, sample handling and sensitivity of the method remain issues to be overcome. It is important to consider proteins, not just genes (DNA or RNA), when analyzing the effect of a substance on humans, because (1) protein levels and activities can differ significantly from RNA levels, (2) many functions of genes can be readily studied using only biochemical techniques, and (3) much regulation of gene activity (including protein stability, modification, and localization) occurs at the protein level (Michaud and Snyder, 2002).

Proteomics dates back to the 1970s, when a newly developed technique of two-dimensional gel electrophoresis (2D GE) on polyacrylamide gels (PAG) came into existence to create databases for all expressed proteins based on differences in isoelectric points and masses (O'Farrell, 1975). In the first dimension, proteins are isoelectrically focused in a pH gradient, allowing for separation based on isoelectric points. In the second dimension, proteins are separated in a mass resolving gel. Semiaccurate determinations for member proteins are determined by staining the gel with protein-binding dyes (Figure 10.1). Some of the problems of the limited dynamic range of silver and Coomassie blue stains have been overcome by labeling with fluorescent dyes (Steinberg et al., 1996). However, this time-consuming, labor-intensive approach has three main problems: (1) the limited resolution of the gel, (2) proteins not readily soluble in aqueous media are rarely detected with 2D GE (Molloy, 2000), and (3) the identity of expressed protein remains unknown (Michaud and Snyder, 2002).

In the 1990s biological mass spectrometry coupled with the availability of the entire human coding sequence in public databases allowed functional analysis of gene products, thus facilitating the development of proteomics (Pandey and Mann, 2000). The coupling of 2D GE to mass spectrometry (MS) provided for both precise mass determination and protein identification (Eng et al., 1994).

Accurate mass determination is best obtained when mass mapping fingerprinting (PMF) is coupled to matrix-assisted laser desorption ionization (MALDI) or electrospray ionization (ESI) time of flight (TOF) MS (Ahmed, 2001; Henzel et al., 1993). Typically, a protein sample is cut from a gel, digested with a specific protease, such as

FIGURE 10.1. 2D gel with *Escherichia coli* lysate. Sample was separated by isoelectric focusing in the first dimension on an IPG, pH 4-7 gradient followed by electrophoresis in the second dimension by 12 percent SDS-PAGE. The protein bands were visualized by silver staining. (*Source:* EMD Biosciences-Novagen, 2002. Reprinted with permission.)

trypsin, and the resulting peptides are accurately sized by MS and compared to predicted sizes of all proteins in the databases. The high-resolution mass determination of the peptides allows accurate identification if several peptides closely match the expected peptide profile of proteins in a database. More recently, the use of tandem mass spectrometry became popular in protein identification (Goodlet and Yi, 2002), as shown in Figure 10.2. MS-MS involves the separation of a peptide mixture in the first MS. Individual peptides are isolated in the MS and fragmented at their bonds. The fragments are then sized in the second MS. The fragmentation pattern allows determination of the peptide sequence, which, when compared to those predicted in the database, allows for the direct identification of a protein.

If the full-length sequences are not available, then electrospray, particularly nanoelectrospray, is used to generate additional partial sequence information to complete the mass information (Wilm and Mann, 1996). Both MALDI-TOF and electrospray can detect low levels of proteins and are suitable for automation. Although of lower throughput, electrospray methods offer the additional information of partial sequence as well as peptide mass, and are particularly useful for posttranslational modifications (Betts et al., 1997). Both methods are often used in a hierarchical approach as shown in Figure 10.3.

Liquid chromatography (LC) methods—in which separation columns are linked directly to MS—allow protein mixtures or a mixture of tryptic digests to be directly separated in the columns and fractions injected into the MS for separation (McDonald and Yates, 2000). One such approach involves peptide separation by LC-LC and peptide

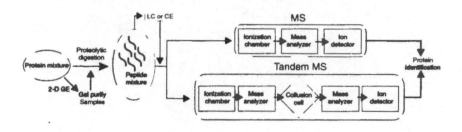

FIGURE 10.2. A scheme outlining uses of mass spectrometer methods for protein identification. A protein mixture is treated with trypsin to fragment the protein into peptides. The resulting peptides can be subjected to either MS or MS-MS for protein identification. (*Source:* From Michaud and Snyder, 2002, with permission.)

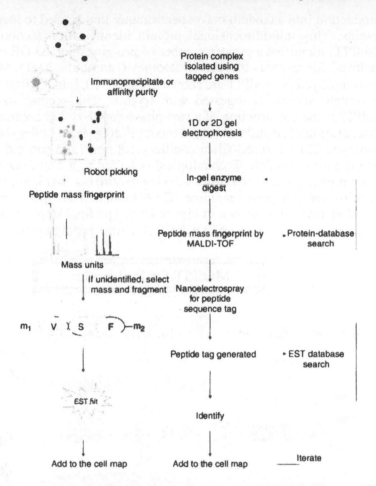

FIGURE 10.3. Mass spectrometric analysis of proteins lacking complete sequence information. Genes of interest are tagged, transfected into cells, and the proteins associated with the cognate-tagged protein are purified by affinity methods. Separation of the protein complex is carried out by one-dimensional or two-dimensional gel electrophoresis. A hierarchical mass spectrometry approach using low-cost and high-throughput methods (MALDI-TOF) is used for initial peptide mass fingerprinting. For completely sequenced genomes, this alone may be sufficient to identify the proteins in the complex. When necessary, electrospray methods are then used to generate peptide-sequence tags for searching protein and EST databases as described by Wilm and Mann (1996). (*Source:* Reprinted from *Trends in Biotechnology,* Vol. 17, Blackstock and Weir, Proteomics: Quantitative and physical mapping of cellular proteins, pp. 121-127, Copyright 1999, with permission from Elsevier.)

introduction into a tandem mass spectrometer that is used to identify proteins. This multidimensional protein identification technology (MudPIT) identified a greater number of proteins than 2D GE when employed for analysis of yeast proteomes (Lin et al., 2001). Major steps employed in MudPIT are shown in Figure 10.4. In the first step, the protein sample is digested with trypsin. The second step of MudPIT is the construction of a two-phase capillary LC column for 2D separation of peptides. Using a model P-2000 laser puller (Sutter Instrument Co.), a fused-silica capillary column (363 µm o.d. and 100 µm i.d.; Polymicro Technologies) is pulled to a 5 µm opening. The column is packed on a custom-made pressure bomb (Figure 10.5). The third step is to perform online 2D LC tandem mass spectrometry (MS-MS) analysis as shown in Figure 10.6. The final step is to use a computer algorithm (e.g., SEQUEST; Thermo Finnigan) to match

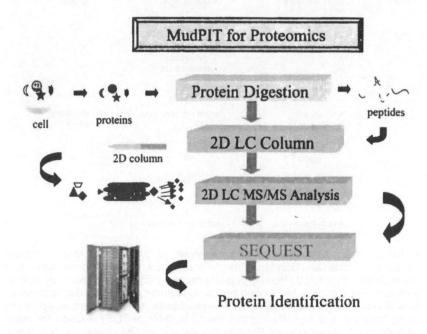

FIGURE 10.4. Major steps in MudPIT. Protein samples (e.g., a cell lysate) are digested with trypsin. The resulting peptides are then loaded onto a 2D micro-capillary column. Using sequential LC steps, peptides are separated and electrosprayed into a mass spectrometer for MS-MS analysis. The MS data are matched to a protein database using the SEQUEST program for protein identification. (*Source:* From Lin et al., 2001, with permission.)

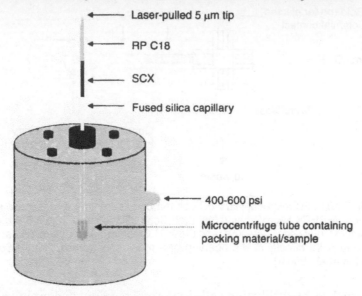

FIGURE 10.5. Preparation of a 2D microcapillary liquid chromatography column using a pressure bomb. The microcapillary column is pulled to yield a tip with a 5 μm opening. The packing materials are suspended in methanol in a microcentrifuge tube, and the tube is placed in the center of the bomb. The packing is achieved by driving the materials into the column under 400-600 psi pressure. The two-phase materials are typically an RP C18 and SCX material. The peptide sample can be loaded onto the column with the same apparatus. (*Source:* From Lin et al., 2001.)

the acquired tandem mass spectra with those predicted from a corresponding sequence database (Link et al, 1999; McDonald and Yates, 2000). This process, which takes several hours to one day, allows for rapid analysis of protein mixtures. A problem with these approaches is that not all proteins present in a mixture are quantitatively separated in peaks. Thus, although they are useful for determining which proteins are present, the relative abundance of the proteins may not be discerned.

To circumvent this problem, several groups have developed advanced MS methods for accurately determining relative amounts of proteins in two samples (e.g., proteins isolated from cells grown in two growth conditions, or in normal versus abnormal condition). This involves labeling protein from each source with a different iso-

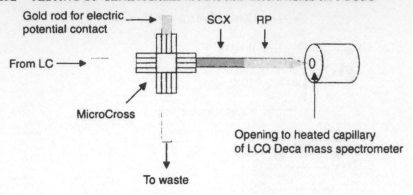

FIGURE 10.6. 2D microcapillary LC column coupled online with a mass spectrometer. A PEEK MicroCross (Upchurch Scientific) is used to join the electric contact, LC inlet line, waste line, and the 2D LC column. The column is carefully placed to point directly at the opening of a mass spectrometer inlet. (*Source:* From Lin et al., 2001.)

tope, such as by incubating cells in normal medium containing either an N^{14} or N^{15} nitrogen source (Oda et al., 1999). Peptide mixtures from two samples are mixed and subjected to LC-MS-MS. Proteins that differ in their two mixtures can be identified rapidly and sequenced. Another approach labels isolated proteins in vitro with either an isotope-coded affinity tag or a solid-phase isotope tag (Gygi et al., 1999; Zhou et al., 2002). Cysteine-containing proteins are specifically labeled with a mass tag that contains seven hydrogen or seven deuterium atoms. By linking the mass tag to either biotin or a solid-phase support, the labeled proteins can be purified, thus simplifying the analysis. Proteins are then subjected to LC-MS-MS for analysis. These approaches show considerable promise for relative protein expression analysis (Michaud and Snyder, 2002).

A protein-chip approach, similar to a DNA microarray, that allows for global analysis of transcriptomics has been developed in which a high-precision robot spots proteins onto chemically derivatized glass slides at extremely high spatial densities. The protein attaches covalently to the slide surface, and can interact specifically with other molecules in a solution labeled with fluorescent dyes (MacBeath and Schreiber, 2000). Antibodies can also be immobilized in an array format onto specially treated surfaces. The surface is then probed with the sample of interest, and only the proteins that bind to the relevant antibodies remain bound to the chip (Lueking et al., 1999). This approach resembles a large-scale version of an enzyme-linked immunosorbed sandwich assay (ELISA). Another approach couples the protein chip with a MALDI readout of the bound material (Nelson, 1997), or with time-of-flight mass spectrometry (James, 2002). However, with all these techniques, the antibodies or proteins bound to chips must be validated for specificity, cross-reactivity, and quantitation to avoid false positive results.

Surface plasmon resonance (SPR) technology has been applied to functional protein studies using the Biacore AB system. As with nucleic acids, proteins bind to target molecules immobilized on the sensor chip and can be captured from complex mixtures with little or no prior purification. The system can detect the specific binding of sub-femtomoles of proteins isolated from complex fluids and molecules as small as 100 Da (Nelson et al., 1999). The SPR system can also supply functional information about the binding event, in addition to determining real-time kinetics (i.e., association and dissociation reaction rates), affinity, and specificity data. SPR technology can also function as a micropurification and recovery system for further analytical studies using mass spectrometry (Lofäs, 2002).

Despite the merits of the various proteomic techniques described earlier, they are limited by their inability to detect all available proteins because the dynamic range of protein expression in a cell is approximately 10^7 and in fluid 10^{12}. Current protein expression methods are sensitive within four orders of magnitude. Thus, at present, no available analytical tool can cover the range of expression of all proteins (James, 2002). The techniques described earlier are still in the developmental process for application to GMO detection, but will ultimately be widely used because of their power of analysis and quantitative ability.

GREEN FLUORESCENT PROTEIN AS A MARKER IN TRANSGENIC PLANTS

The use of transgenic plants in agriculture raises concern about the flow of transgenes to wild relatives (Raybould and Gray, 1993). Moreover, the in-field monitoring of transgene expression can be a cause of concern if the underexpression of an herbicide-tolerant gene is detected in crop plants (Harper et al., 1999). The potential of transgene escape has led to development of a simple gene monitoring system that is applicable to field use. Detection of antibiotic- and herbicide-resistant genes is unsuitable, because it requires destructive plant tissue sampling or time-consuming assays. A desirable system should allow detection of transgenes and their expression in real time and online in plants (Harper et al., 1999).

A green fluorescent protein (GFP) from the jellyfish *Aequorea victoria* would be a suitable in vivo marker because it fluoresces green (distinguishable from wild-type plants that autofluorece reddish purple) when excited with ultraviolet (UV) or blue light without the addition of substrates or cofactors. This property makes GFP useful for monitoring transgenes for agricultural and ecological applications (Stewart, 1996) since autofluorescent plants can be detected with an appropriate wavelength of light or under a fluorescent microscope. The gene for GFP has been cloned (Prasher et al., 1992) and modified for increased expression and improved fluorescence in plants (Reichel et al., 1996). The *mGFPer* gene targeted to the endoplasmic reticulum has been modified for cryptic intron missplicing and codon usage. Two mutations, V163A and S175G, enhance the folding of this gene at high temperature, and an I167T substitution changes the UV (λ 395 nm) and blue light (λ 473 nm) maxima of the *mGFP5er* gene to equal amplitudes (Siemering, 1996). A practical way to detect transgenes in the field has been to monitor the presence of expression of an "agronomically important" gene by linking it to a marker gene such as GFP. The GFP fluorescence can indicate the expression of *Bacillus thuringiensis* gene *Cry1Ac* when introduced into tobacco and oilseed rape, as demonstrated by insect bioassays and Western blot analysis (Harper et al., 1999). Limitations to the GFP monitoring system include (1) the possibility that transgenes unlink over multiple generations, and (2) differences in relative expression of the two transgenes, unless both genes are integrated in a transcriptionally active area of the gene, in which case they are both expressed to a high degree (Allen et al., 1996). Host plants synthesizing GFP in the field were reported not to suffer a "fitness cost" (i.e., no toxicity to tobacco plants was observed) (Harper et al., 1999).

CONCERNS ABOUT GENE FLOW,
LIABILITIES, REGULATORY CLIMATES,
AND INTELLECTUAL PROPERTY RIGHTS

As mentioned earlier, gene flow from crops to wild or weedy relatives has often been cited as a potential risk in commercialization of transgenic crops (NRC, 2002). Although crops and weeds have been exchanging genes for centuries, biotechnology raises additional concerns because it allows novel genes to be introduced into many diverse crops, each with its own potential to outcross. Moreover, it is currently impossible to prevent gene flow between sexually compatible species in the same area (Snow, 2002). Questions about the ecological and agronomic consequences of gene flow are largely unanswered because this area of research has not received appropriate funding, and thus not much research has been conducted on the extent of these risks. In the absence of these studies, it is unwise and simplistic to claim that gene flow poses no—or negligible—risks to agricultural biodiversity (Abbott et al., 1999). A growing concern threatening current production practices in intensive agriculture is the cross pollination of GMO crop varieties with conventional varieties. Another concern is the germination of volunteer seeds (e.g., seeds dropped, blown, or inadvertently planted). Fortunately, the growth of volunteer seeds is controllable by chemicals, but at an added cost for producers (Smyth et al., 2002), which leads some individuals/organizations to question the economy of using herbicide-tolerant varieties (Benbrook, 2001).

It appears that insufficient attention—as a result of the introduction of GMOs to plants—has been given to the following three issues: (1) the introduction of the same gene into different types of cells can produce distinct proteins; (2) the introduction of any gene (either from the same or different species) can significantly change overall gene expression and, thus, the phenotype of the recipient cells; and (3) enzymatic pathways introduced to synthesize micronutrients may interact with endogenous pathways leading to production of novel, metabolically active molecules. As with secondary modifications, it is possible that any, or all, of the above perturbations result in unpredictable outcome (Schubert, 2002). Although short- and long-term toxicity and metabolic studies could be carried out to address concerns, it is probably not feasible that they will detect relevant changes

· unless extensive safety testing is carried out on GM crops. More important, it is essential to know the effect(s) we wish to monitor so that they can be targeted and carefully studied.

The first generation of "input traits" of GM crops (e.g., traits with purely agronomic benefits) is entering its ninth year, with the majority derived from these leading crops in North America, namely canola (rapeseed), corn, and soybean. These products harbor traits that serve an agronomic purpose (i.e., benefiting farmers, but not necessarily consumers). They entered the North American market with minimum regulations and without segregation, and have been judged by regulators as substantially equivalent to existing varieties. Second-generation crops, which involve output modifications (traits with health and nutritional benefits) (Agius et al., 2003), are unlikely to be cleared unless their purity is assured. This is a problematic prospect given the current difficulties of attaining gene containment. Third-generation crops with new industrial, nutraceutical, or pharmaceutical properties will most likely require an effective gene control system for them to be allowed to enter the markets (Kleter et al., 2001).

Regardless of how effective regulations are, some producers (either deliberately or inadvertently) will misappropriate these technologies, creating risks and liabilities. Moreover, many plant species are sexually promiscuous, creating natural gene flow to related species, and leading to the following liability issues: (1) the potential of volunteer seeds inadvertently left in the field to germinate the following year(s), (2) the potential for pollen flown from GM crops to non-GM crops, (3) the potential for comingling of GM and non-GM crops, which could jeopardize the value of both crops and product lines if transgenes remain undetected before processing, and (4) the potential for environmental risks associated with uncontrolled gene flow from GM varieties into related plants, which will impede export of GM varieties to countries not willing to adopt the new technologies. These liability issues have resulted in some disastrous consequences and imposed significant cost on the food industry (Smyth et al., 2002).

About 800 million people (18 percent of the world's population) living in developing countries are malnourished, primarily because of poverty or unemployment, of which about 6 million children under five years old die annually from malnutrition. It is estimated that about 650 million of the poorest people live in rural areas in developing countries, where the local production of foods on small farms is the main economic activity (Royal Society, 2000). Deficiencies in micronutrients (especially vitamin A, iodine, and iron) are widespread in the developing countries. Furthermore, changes in the pattern of global climate and alterations of land use will exacerbate the problem of regional production and demand for food. It is not expected that cultivation area will increase, as most land suited for agriculture is already in use. Thus, advances in food production, distribution, and access are expected to come primarily from GM and related technologies (Serageldin, 1999).

In the 1950s and the 1960s, the Green Revolution was carried out in publicly supported research institutions and was funded by the public sector and charitable foundations to ensure that research of relevance to small-scale farmers and complex tropical and subtropical environments was conducted. This revolution in agriculture produced many plant varieties and dramatically increased the yield of traditional crops in developing countries, especially in the Indian subcontinent and China (Borgstrom, 1973). Despite past successes, the rate of increase of crop production has dropped (yield increase of 3 percent per year in the 1970s declined to 1 percent per year in the 1990s) (Conway, 1999). Since the 1990s, the balance of funding for this research has shifted significantly from the public to the private sector, subjecting research priorities to market forces and gearing them to the need of large-scale commercial agriculture enterprises (who have also been granted broad patents related to GMOs for minor modifications of gene sequences), rather than to those of subsistence farmers (Royal Society, 2002).

The biotechnology revolution plays an important role in the problem solving of food security concerns, poverty reduction, and environmental conservation issues, particularly in developing countries, but questions related to ethics, intellectual property rights, and biosafety issues must be addressed and solved. At present, it appears that many less developed countries are reluctant to join in international intellectual property agreements on plants because they believe that such agreements will create a system that strongly favors the corporate sector, while simultaneously hampering the public and private sector efforts that support their own citizens. Ways must be found to overcome the adverse effects of patents and intellectual property policies on the conduct of research and the exchange, use, and improvement of germplasm. Fairness issues also must be dealt with. The general public and world community have come to accept great disparity in wealth and income, but the same cannot be said about access to and control over seeds and genetic varieties (Benbrook, 2000).

Patenting issues have two sides. Supporters stress that if the private sector is to invest large sums of money in agrobiotechnology R&D, it must recoup its investments (http://www.bio.org). Opponents argue that patenting will lead to monopolization of knowledge, restrict access to germplasm, controls research focus and direction, and increase the marginalization of the majority of world population (http://www.rafi.org) (Royal Society, 2002).

To ensure hefty financial return for their investment, many biotechnology seed companies have sought to prevent the use of second-generation seed produced from transgenic crops by employing genetic use restriction technology (GURT), or terminator technology (Visser et al., 2001). This technology involves the use of chemical treatment of seeds or plants that either inhibits or activates specific genes involved in germination. The possible commercialization of GURT technology has generated considerable public debate from growers and agencies representing developing countries, governments, environmental groups, and other organizations. Concerns have been expressed that terminator technology could threaten landrace varieties, increase corporate concentration, reduce biological diversity, and ultimately destabilize the agroeconomics of less-developed countries (Visser et al., 2001).

In an alternative GURT, the transgenic trait would be expressed only if a certain chemical activator was applied to seeds or plants. In this case, farmers would retain the ability to save their own seed, yet lack access to the added traits in the absence of payment for chemical activator(s). However, GURTs may have beneficial applications for growers, consumers, and the environment that should not be entirely overlooked in debates over intellectual property rights. For example, GURTs could be used to prevent transgenes from spreading to closely related wild plants by preventing germination of any crossbred seeds. This technology could potentially eliminate the problems of "volunteer" plants that appear from seed left in the field after harvest. Volunteer plants must be eliminated before the next crop is planted because they are hosts for pests and pathogens. GURTs can prevent the escape or spread of potentially harmful traits (e.g., herbicide tolerance) from the GM crops. Moreover, they can reduce product liability assigned to seed growers by preventing contamination through commingling with non-GM crops (Smyth et al., 2002). If GURTs are not used, other central mechanisms must be instituted. Although the initial cost of introducing control mechanisms may be high, the long-term benefit of such technologies may justify their adoption. Ultimately, the inability to manage risks and control liabilities can reduce net returns or investments to the extent that GM technology may become unattractive. Capital is an important liquid commodity in today's market place, and by prohibiting the commercialization of relatively safe GM products, countries doing so risk losing—not only investment capital, but also R&D-intensive farming (Smyth et al., 2002).

International harmonization of legislation on GM-derived products would be advantageous for everyone concerned (e.g., manufacturers, users, and consumers worldwide) and would prevent needless trade barriers. Efforts are being made at the international level with the Food and Agriculture Organization/World Health Organization (FAO/WHO) *Codex Alimentarius,* however, the process may take several years to achieve (FAO/WHO, 2001). The "Cartagena Protocol on Biosafety," an agreement stemming from the 1992 United Nations Convention on Biodiversity, was signed in Nairobi, Kenya, in May 2000 by 64 governments and the European Union. The protocol puts into effect rules that govern the trade and transfer of GMOs across international borders, such as labeling of GMO commodity shipments, and allows governments to prohibit the import of GM food when concerns exist over its safety (Gupta, 2000). These universal legislations have made it imperative for governments, the food industry, crop producers, and testing laboratories to develop ways to accurately quantitate GMOs in crops, foods, and food ingredients to ensure compliance with threshold levels of GM products (Ahmed, 2002).

ASSESSMENT OF SAFETY ISSUES

Biotechnology offers a variety of potential benefits and risks. It has enhanced food production by making plants less vulnerable to drought, frost, insects, and viruses, and by enabling plants to compete more effectively against weeds for soil nutrients. In a few cases, it has also improved the quality and nutrition of foods by altering their composition (Ahmed, 2002).

Table 10.1 summarizes the GM foods evaluated by the U.S. Food and Drug Administration (FDA). It shows that the majority (48 out of 52) of modifications have been aimed at increasing crop yield for farmers by engineering a food plant to tolerate herbicides or attacks from insects and viruses. Moreover, only two food plants have been altered to produce modified oil (e.g., soybean and canola plants). The modified soybean produces healthier oil, and the canola plant produce laurate cooking oil. Since soybean oil is the most commonly consumed plant oil in the world, these newly produced oils could significantly improve health for millions of people worldwide (GAO, 2002).

The U.S. accounts for about three-quarters of GM food crops planted globally. For three key crops in the United States (corn, soybean, and cotton), a large number of farmers have chosen to plant GM varieties. In 2001, GM varieties accounted for about 26 percent of the corn, 68 percent of the soybean, and 69 percent of the cotton planted in the United States. These crops are the source of various ingredients used extensively in many processed foods, such as corn syrup, soybean oil, and cottonseed oil, and they are also a major U.S. commodity export (Ahmed, 2002).

TABLE 10.1. GM foods for human consumption evaluated by the FDA

Modified attribute	Insect resistance	Viral resistance	Herbicide tolerance	Modified oil	Plant reproductive sterility	Delayed ripening/ softening
GM plant product— # of plant varieties	Corn—8 Tomato—1 Potato—4 Cotton—2	Squash—2 Papaya—1 Potato—2	Corn—9 Rice—1 Canola—8 Sugar Beet—2 Flax—1 Cotton[a]—4 Radish—1 Soybean—2	Soybean—1 Canola—1	Corn—3 Canola—3 Radish—1	Cantaloupe—1 Tomato—4
Total[b]	15	5	28	2	7	5

Source: From GAO, 2002.
[a]Cotton seed has been used as a protein source in candy.
[b]Fifty products have been evaluated, as of April 2002. The total number of modified attributes is 62 because several products were modified with multiple attributes.

According to reports from the International Food Biotechnology Council (IFBC, 1990), the Organization for Economic Cooperation and Development (OECD, 2001), FAO/WHO (2000), the Royal Society (2002), and the FDA (1992), foods from GM plants pose three risks to human health: they can contain allergens, toxins, or antinutrients. These risks, however, are not unique to GM foods. The concept of substantial equivalence, which was developed in 1993 by OECD as part of a safety evaluation framework, is based on the idea that existing foods can serve as a basis for comparing the properties of GM foods with the appropriate counterpart to identify similarities and differences between existing foods and new products, which are then subjected to further evaluation (OECD, 1993).

Current food safety regulations for traditional food crops are naturally less stringent than those applied to GM foods because of a long history of traditional breeding that has given insight into the beneficial and adverse compounds that have either increased or decreased through breeding. This targeted approach has resulted in relatively safe food, and should continue to be the guiding principle for assessing food products (NRC, 2001).

In the United States, companies that wish to submit new GM foods for FDA evaluation must perform a regimen of tests to obtain safety data. The FDA's (1992) policy on safety assessment of GM foods describes the data the agency recommends it receive to evaluate these foods. Figure 10.7 provides an example of the tests, including an analysis of (1) the source of the transferred genetic material, (2) the degree of similarity between the amino acid sequences in the newly introduced proteins of the GM food and the amino acid sequences in known allergens, toxins and antinutrients, (3) data on in vitro digestibility (i.e., how readily the proteins break down in simulated digestive fluids), (4) the comparative severity of individual allergic reactions to the GM product and its conventional counterpart as measured through blood screening, when the conventional counterpart is known to elicit allergic reactions or allergenicity concerns remain, and (5) data on any changes in nutrient substances, such as vitamins, proteins, fats, fiber, starches, sugars, or minerals due to genetic modifications (GAO, 2002). The tests do not guarantee absolute safety of GM food, but ensure comparable safety. There is no assurance that even conventional foods are completely safe, since some people suffer from

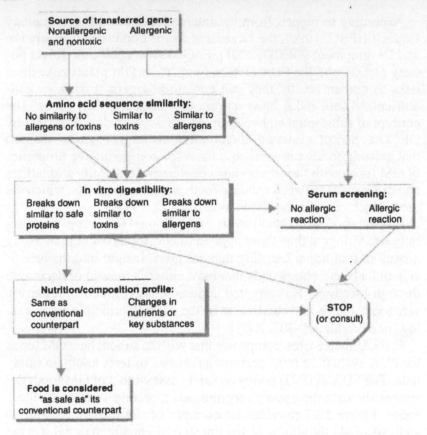

FIGURE 10.7. Example of the regimen used for assessment of the safety of GM foods carried out by the U.S. FDA. (*Source:* Adapted from GAO, 2002.)

allergic reactions, and conventional foods contain toxins and anti-nutrients (Ahmed, 1999).

Although the current regimen of tests seems adequate for safety assessments, limitations exist for individual tests. For example, the acceptability of amino acid sequence similarity test results is limited, in part, because no agreement exists on what level of amino acid similarity indicates a likelihood of allerginicity. Therefore the need for additional testing effectively ensures that the FDA obtains the data necessary for evaluating the potential risks of GM foods. These practices include (1) communicating clearly what safety data are important to the FDA's evaluation of GM food safety, (2) having teams of FDA experts representing diverse disciplines perform the evaluations, and (3) tailoring the level of evaluation to match the characteristics of each GM food. The FDA's management practices constitute internal controls. For GM food, the evaluation process, known as a "consultation," lasts between 18 months and three years, depending on the product. The principles embodied in the FDA's 1992 statement have guided the consultation for the 50 GM foods evaluated, and must include concurrence from every member of the Biotechnology Evaluation Team, which is generally composed of a consumer safety officer (who serves as the project manager), molecular biologist, chemist, environmental scientist, toxicologist, and nutritionist. The FDA varies its level of evaluation based on the degree of novelty of the GM food submission to allow resources to be devoted where they are most needed, thus giving the evaluation team the time to obtain, examine and evaluate necessary safety data. It is expected that future GM foods that include enhancements to their nutritional value will increase the difficulty of assessing their safety, as some of the new ingredients in GM foods will differ significantly from ingredients that have a history of safe use.

New testing technologies are being developed, but have not yet been applied widely to GM foods because of technical limitations, such as unavailability of internal standards or baseline information. Reliability assessments of these technologies are carried out to differentiate between naturally occurring changes due to a wide range of environmental conditions and the effects of deliberate genetic modifications. These technologies, which employ a nontarget approach to increase the chance of detecting unintended effects of genetic modifications (e.g., creation of a toxin), include (1) DNA microarray technologies that identify gene sequences and determine the expression level (or abundance) of the genes, (2) proteomics, which can analyze up to 100,000 proteins simultaneously to provide information on the creation of new (or modified) proteins. Considering that the function of proteins in a plant may constantly change in their secondary, tertiary, or quaternary structure depending on their interaction or expression in different cells and tissues, this change may profoundly influence proteins' electrophgoretic behavior and nuclear mass (Kuiper et al., 2001) and (3) metabolic profiling that can analyze the 2,000 to

3,000 metabolites in people and 3,000 to 5,000 metabolites in plants to determine whether changes in small molecules have occurred (FDA, 1992).

Because no scientific evidence has shown that GM foods cause long-term harm, such as increased cancer rates, there is no plausible hypothesis of harm. These hypotheses are required to determine what problems to search for, test, and potentially measure (Kuiper et al, 2001). Technical challenges make long-term monitoring of GM foods virtually impossible, such as the following:

1. Conducting a long-term monitory requires an experimental group(s) that has consumed GM foods and a control group consisting of people who have not eaten GM foods. In the United States, where labeling is not required, it would nearly be impossible to identify such individuals reliably.
2. Even in countries where GM foods are labeled, it would be very difficult to separate the health effects of GM foods from their conventional counterparts because few nutritional differences exist between these foods. Furthermore, there would be practical challenges in feeding both the experimental and control groups diets that contain large quantities of such GM foods as soybean or corn, as well as their conventional counterparts.
3. Since the long-term human health effects of consuming most foods are not known, there is no baseline information against which to assess health effects caused by GM foods
4. Changes in human food consumption patterns, such as the addition and removal of various foods, add new variables to the diet, and compound the difficulty of conducting long-term epidemiological monitoring (GAO, 2002).

FAO/WHO asserts that very little is known about potential long-term effects of any food, and identification of such effects is further confounded by the great variability in people's reaction to foods, especially those containing allergens. Moreover, epidemiological studies are unlikely to differentiate the health effects of GM foods from the many undesirable effects of conventional foods, which include consumption of cholesterol and fat. Thus, the identification of long-term effects specifically attributable to GM foods seems highly unlikely (FAO/WHO, 2000).

In Europe, a decision to label GM foods to allow consumers an informed choice has been firmly established. In Switzerland, a country that is not a member of the EU, enforcement of the Swiss Food Regulation and the EU Novel Food Regulation is based on a quantitative PCR detection system specific for the cauliflower mosaic virus (CaMV) 35S promoter (Figure 10.8a). Labeling in Switzerland is mandatory if the GMO content is above 1 percent on the basis of ingredients. If the GMO cannot be identified, samples have to be checked for the absence of CaMV, and DNA preparations have to be checked for the absence of PCR inhibitors. Moreover, it is uncertain whether the compilation of a negative list of ingredients containing no detectable GMOs (i.e., highly refined oils, lecithin, certain products derived from starch and cocoa) in countries setting limits is adequate, since consumers are more interested in the authenticity of food rather than analysis in countries that mandated labeling GM foods (Hübner et al., 1999).

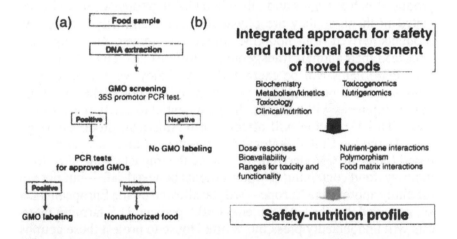

FIGURE 10.8. Examples of GMO testing and safety evaluations: (a) Swiss procedure for GMO testing of food (*Source:* Reprinted from *Food Control,* Vol. 10, Hübner et al., Quantitative competitive PCR for the detection of genetically modified organisms in food, pp. 353-358, Copyright 1999, with permission from Elsevier.); (b) integrated approach for safety evaluation of genetically modified foods (*Source:* Reprinted from *The Plant Journal,* Vol. 27, Kuiper et al., Assessment of the food safety issues related to genetically modified foods, pp. 503-528, Copyright 2001, with permission from Blackwell Publishing Ltd.)

In Europe, as in the United States, the assessment of GM plants and foods with enhanced nutritional properties is believed to benefit from a balanced focus on the simultaneous characterization of inherent toxicological risks and nutritional benefits requiring an integrated multidisciplinary approach that incorporates molecular biology, genetics, toxicology, and nutrition (Figure 10.8b). Issues that must be considered include (1) evidence for nutritional/health claims and target population, (2) toxicological dose ranges of selected compounds, (3) impact on overall dietary intake and associated effects on consumers, (4) interactions between food constituents and food matrixes, and (5) postmarket surveillance of introduced products. Thus far, the application of the principle of substantial equivalence for assessing the safety of foods containing GMOs has proven adequate, and it does not appear that alternative safety assessment strategies will replace it (Kuiper et al., 2001).

A new, strict legislation agreed upon in November 2002 between ministers of individual EU member states extends labeling to end-products such as sugars and oils when GM ingredients cannot be detected in them, as they are physically and chemically identical to products derived from non-GM sources. Moreover, meat suppliers who feed their animals transgenic grains must also label their products. Food items will be exempted only if they were derived from crop material containing <0.9 percent GMO. A comprehensive tracing of shipments will undoubtedly be essential to verify this requirement. This legislation will affect North American producers (e.g., Canada and the United States) because most of their soy- and corn-based foods are GM-derived. To reduce the impact on the U.S. food industry, food containing GM ingredients believed safe—but not yet officially approved in Europe—will be allowed on the European market provided that the GM content is <0.5 percent. U.S farming interests will undoubtedly press the White House to protest these actions to the World Trade Organization to prevent barriers on importing their GM foods, which is believed to cost U.S. corn producers $250 millions per year in lost sales. Moreover, the U.S. industry is also quite worried about the effect the European policy will have on importers in developing countries. For example, in October 2002, Zambia refused 63,000 tons of GM corn from the U.S. intended to help relieve the current famine in southern Africa, fearing that this corn will contaminate Zambia's agriculture commodities and make them unmarketable in Europe (Mitchell, 2003).

CONCLUSIONS

Table 10.2 summarizes the most commonly used tests employed for the detection of GMOs in foods and ingredients with respect to ease of use, requirement for specialized equipment, assay sensitivity, time to run, quantitation, cost, suitability for routine field testing, or utility as research tools. To respond to regulations requiring food labeling, such as those of the EU, a tiered approach might be employed, starting with qualitative PCR for GMO detection. If no GMOs were detected with a validated qualitative method, the product(s) would be evaluated for the presence of protein. If no protein is detected, the product is presumed undetectable. If the qualitative PCR shows a positive result, the product is considered to be a "nonapproved GMO," and a validated quantitative (Q)-PCR or real-time-PCR is used to detect the level of GMO. If the GMO level is above an established threshold, the product is labeled as "nonapproved GMO," but if below the threshold, the product need not be labeled (Kuiper, 1999). Q- or real-time-PCR might best be applied at early stages in the food chain. The high sensitivity and specificity of quantitative PCR methods and their flexibility with different food matrixes make them suitable for detecting GMOs at low thresholds in various foods (Ahmed, 2002).

The greatest uncertainty of using DNA-based assays, such as for protein-based methods, is that not all products derived from GM foods (e.g., refined oil) contain significant amounts of DNA. In addition, heating, high pH, and other processes associated with food production can degrade DNA. Similarly, if GMOs are expressed on a relative basis (i.e., percent GMO), it is important to know whether the estimate is based on total DNA from all sources or on the basis of analyzed product DNA. This pragmatic approach, known as "genetic equivalence," was correlated with results of studies in which GMO content was expressed as a percentage of mass. Using the genetic equivalent approach to assess the GMO content of food ingredients, in addition to tracing the ingredients used, should allow for an accurate estimation of GMOs. This approach is also consistent with cur-

TABLE 10.2. Comparison among tests commonly used for detection of GMOs in foods[a]

Parameter	Protein-based			DNA-based					
	Western blot	ELISA	Lateral flow strip	Southern blot	Qualitative[a,b] PCR	QC-PCR and limiting dilution	Real-time PCR	DNA Micro-arrays	DNA sensors
	Difficult	Moderate	Simple	Difficult	Difficult	Difficult	Difficult	Difficult	Difficult
Needs special equipment	Yes	Yes	No	Yes	Yes	Yes	Yes	Yes	Yes
Sensitivity	High	High	High	Moderate	Very high	High	High	High	Low
Duration[c]	2 d	30-90 min	10 min	6 h[d]	1.5 d	2 d	1 d	2 d	2 d
Cost/sample	US$150	US$5	US$2	US$150	US$250	US$350	US$450	US$600	US$200
Provides quantitative results	No	Yes[e]	No	No	Yes	Yes	Yes[f]	Yes	No
Suitable for field test	No	Yes[e]	Yes	No	No	No	No	Yes	Yes
Employed mainly in	Academic labs	Test facility	Field testing	Academic labs	Test facility	Test facility	Test facility	Academic[c,g] labs	Academic[c,g] labs

(Source: Modified from Ahmed, 2002.)

aNIR detects structural changes (not DNA or protein), is fast (<1 min) and inexpensive (~$1).
bIncluding nested PCR and GMO Chip
cExcluding time allotted for sample preparation
dWhen nonradioactive probes are used; otherwise 30 h with ^{32}P-labeled probes
eAs in the antibody-coated tube format
fWith high precision
gIn development

rent EU food-labeling regulations that focus on ingredients, and is applicable to finished products containing more than one GMO-derived ingredient (Kuiper, 1999).

New, high-throughput parallel technologies, such as DNA microarrays and proteomics, will increasingly find their niches in the safety assessment chain of tests required for evaluating GMOs, but only after they are standardized and fully developed and their costs are brought to an affordable level.

REFERENCES

Abbott, A., Dickson, D., and Saegusa, A. (1999). Long-term effects of GM crops serves up food for thought. *Nature* 398:651-655.

Agius, F., González-Lamothe, R., Cabarello, J.L., Mañoz-Blanco, J., Botella, M.A., and Valpuesta, V. (2003). Engineering increased vitamin C levels in plants by overexpression of a D-galacturonic acid reductase. *Nature Biotechnol.* 21:177-181.

Ahmed, F.E. (1999). Safety standards for environmental contaminants in foods. In *Environmental Contaminants in Food* (pp. 550-570), C. Moffat and K.J. Whittle, eds. Sheffield Academic Press, United Kingdom.

Ahmed, F.E. (2001). Analysis of pesticides and their metabolites in foods and drinks. *Trends Anal. Chem.* 22:649-655.

Ahmed, F.E. (2002). Detection of genetically modified organisms in foods. *Trends Biotechnol.* 20:215-223.

Allen, G.C., Hall, G. Jr., Michalowski, S., Newman, W., Spiker, S.,Weissinger, A.K., and Thompson, W.F. (1996). High-level transgenic expression in plant cells: Effects of a strong scaffold attachment region from tobacco. *Plant Cell* 8: 899-913.

Benbrook, C. (2000). *Who controls and who will benefit from plant genomics?* Paper presented at the 2000 Genome Seminar: Genomic Revolution in the Fields: Facing the Needs of the New Millennium, AAAS Annual Meeting, February 19, Washington, DC.

Benbrook, C. (2001). Do GM crops mean less pesticide use? *Pesticide Outlook* (October):204-207.

Betts, J.C., Blackstock, W.P., Ward, M.A., and Anderson, B.H. (1997). Identification of phosphorylation site on neurofilament protein by nanoelectrospray mass spectrometry. *J. Biol. Chem.* 272:12922-12927.

Blackstock, W. and Weir, M.P. (1999). Proteomics: Quantitative and physical mapping of cellular proteins. *Trends. Biotechnol.* 17:121-127.

Borgstrom, G. (1973). *The Food and People Dilemma.* Danbury, North Sectuate, MA.

Charlton, A., Allnutt, T., Holmes, S., Chisholm, J., Bean, S., Ellis, N., Mullineaux, P., and Oehlschlager, S. (2004). NMR profiling of transgenic peas. *Plant Biotechnol.* (in press).

Conway, G. (1999). *The Doubly Green Revolution: Food for All in the 21st Century.* Penguin Books, London, United Kingdom.

EMD Biosciences-Nouagen (2002). Proteomics I: Tools for protein purification and analysis. *Novagen Catalog,* 2001-2002. Novagen, Madison, WI.

Eng, J.K., McCormack, A.L., and Yates, J.R. III (1994). An approach to consolidate tandem mass spectral data of peptides with amino acid sequences in protein database. *J. Am. Soc. Mass Spectrom.* 5:976-989.

Food and Agriculture Organization/World Health Organization (FAO/WHO) (2000). *Safety Aspects of Genetically Modified Foods of Plant Origin.* Report of a Joint FAO/WHO Expert Consultation on Foods Derived from Biotechnology, May 29-June 2, Geneva, Switzerland.

Food and Agriculture Organization/World Health Organization (FAO/WHO) (2001). *Joint FAO/WHO Food Standard Programme.* Codex Ad Hoc Task Force on Foods Derived from Biotechnology, Second Session, Chiba, Japan, March 25-29. Codex Alimentarius Commission, FAO of the United Nations, Rome, Italy.

Food and Drug Administration (FDA) (1992). Statement of policy: Foods derived from new plant varieties. *Federal Register* 57:22984-23005 (May 29).

General Accounting Office (GAO) (2002). *Experts View Regimen of Safety Tests As Adequate, but FDA's Evaluation Process Could be Enhanced.* Report to Congressional requesters. U.S. General Accounting Office, May.

Goodlett, D.R. and Yi, E.C. (2002). Proteomics without polyacrylamide: Qualitative and quantitative use of tandem mass spectrometry in proteome analysis. *Funct. Integr. Genomics* 2:138-153.

Gupta, J.A. (2000). Governing trade in genetically modified organisms: The Cartagena Protocol on Biosafety. *Environment* 42:22-23.

Gygi, S.P., Rist, B., Gerber, S.A., Turecek, F., Gelb, M.H., and Aebersold, R. (1999). Quantitative analysis of complex protein mixtures using isotope-coded affinity tag. *Nat. Biotechnol.* 17:994-999.

Hall, R., Beale, M., Fiehn, O., Hardy, N., Sumner, L., and Bino, R. (2002). Plant metabolomics: The missing link in functional genomics strategies. *Plant Cell* 14:1437-1440.

Harper, B.K., Mabon, S.A., Leffel, S.M., Halfhill, M.D., Richards, H.A., Moyer, K.A., and Stewart, C.N. Jr. (1999). Green fluorescence protein as a marker for expression of a second gene in transgenic plants. *Nature Biotechnol.* 17:1125-1129.

Henzel, W.J., Billeu, T.M., Stults, J.T., Wong, S.C., and Grimley, C. (1993). Identifying proteins from two-dimensional gels by molecular mass searching of peptide fragment in protein sequence database. *Proc. Natl. Acad. Sci. USA* 93:5011-5015.

Hübner, P., Studer, E., and Lüthy, J. (1999a). Quantitation of genetically modified organisms in food. *Nature Biotechnol.* 17:1137-1138.

Hübner, P., Studer, E., and Lüthy, J. (1999b). Quantitative competitive PCR for the detection of genetically modified organisms in food. *Food Control* 10:353-358.

International Food Biotechnology Council (IFBC) (1990). Biotechnologies and foods assuring the safety of foods produced by genetic modification. *Regul. Toxicol. Pharmacol.* 12:S1-S196.

James, P. (2002). Chips for proteomics: A new tool or just hype? *BioTechniques* 33: S4-S13.

Kleter, G.A., van der Krieken, W.M., Kok, E.J., Bosch, D., Jordi, W., and Gilissen, L.J.W.J. (2001). Regulation and exploitation of genetically modified crops. *Nature Biotechnol.* 19:1105-1110.

Kuiper, H.A. (1999). Summary report of the ILSI Europe workshop on detection methods for novel foods derived from genetically modified organisms. *Food Control* 10:339-349.

Kuiper, H.A., Kleter, G., Noteborn, H.P.J.M., and Kok, E. (2001) Assessment of the food safety issues related to genetically modified foods. *Plant J.* 27:503-528.

Kuiper, H.A., Kok, E.J., and Engel, K-H. (2003). Exploitation of molecular profiling technique for GMO food safety assessment. *Curr. Opinion Biotechnol.* 14:238-243.

Lin, D., Alpert, A.J., and Yates, J.R. III (2001). Multidimensional protein identification technology as an effective tool for proteomics. *American Genomic/Proteomic Technol.* (July/August):38-47. Figures reprinted with permission from the Genomic/Proteomic Laboratory.

Link, A.J., Eng, J., Schieltz, D.M., Carmack, E., Mize, G.J., Morris, D.R., Garvik, B.M., and Yates, J.R. III (1999). Direct analysis of protein complexes using mass spectrometry. *Nature Biotechnol.* 17:676-682.

Lofås, S. (2002). Finding function for proteomics. *Genetic/Proteomic Technol.* (Nov/Dec):33-36.

Lueking, A., Horn, M., Eickhoff, H., Lehrach, H., and Walter, G. (1999). Protein microarrays for gene expression and antibody screening. *Anal. Biochem.* 270: 103-111.

MacBeath, G. and Schreiber, S.L. (2000). Printing proteins as microarrays for high-throughput function determination. *Science* 289:1760-1762.

McDonald, W.H. and Yates, J.R. III (2000). Proteomic tools for cell biology. *Traffic* 1:747-754.

Michaud, G.A. and Snyder, M. (2002). Proteomic approaches for the global analysis of proteins. *BioTechniques* 33:1308-1316.

Mitchell, P. (2003). Europe angers US with strict GM labeling. *Nature Biotechnol.* 21:6.

Molloy, M.P. (2000). Two-dimensional electrophoresis of membrane proteins using immobilized pH gradients. *Anal. Biochem.* 280:1-10.

National Research Council (NRC) (2001). *Genetically Modified Pest-Protected Plants: Science and Regulation.* National Academy Press, Washington, DC.

National Research Council (NRC) (2002). *Environmental Effects of Transgenic Plants*. National Academy Press, Washington, DC.

Nelson, R.W. (1997). The use of bioreactive probes in protein characterization. *Mass Spectrom. Rev.* 16:353-376.

Nelson, R.W., Jarvik, J.W., Tallon, B.E., and Tubbs, K.A. (1999). BIA/MS of epitope tagged peptides directly from *E. coli* lysate: Multiplex detection and protein identification at low-semtomole to subfemtomole levels. *Anal. Chem.* 71: 2858-2865.

Oda, Y.K., Huang, K., Cross, F.R., Cowburn, D., and Chait, B.T. (1999). Accurate quantitation of protein expression and site-specific phosphorylation. *Proc. Natl. Acad. Sci. USA* 86:6591-6596.

O'Farrell, P. (1975). High resolution two-dimensional electrophoresis of proteins. *J. Biol. Chem.* 250:4007-4021.

Organization for Economic Cooperation and Development (OECD) (1993). *Safety Evaluation of Foods Derived by Modern Biotechnology*. Paris, France.

Organization for Economic Cooperation and Development (OECD) (2001). Report of the OECD Workshop on Nutritional Assessment of Novel Foods and Feeds, Ottawa, Canada, February 5-7. Paris, France.

Pandey, A. and Mann, M. (2000). Proteomics to study genes and genomes. *Nature* 405:837-846.

Prasher, D.C., Eckenrode, V.K., Ward, W.W., Prendergast, F.G., and Cormier, M.J. (1992). Primer structure of the *Aequorea victoria* green fluorescent protein. *Gene* 111:229-233.

Raybould, A.F. and Gray, A.J. (1993). Genetically modified crops and hybridization with wild relatives: A UK perspective. *J. Appl. Ecol.* 30:199-219.

Reichel, C., Mathur, J., Eckes, P., Langenkemper, K., Koncz, C., Schell, J., Reiss, B., and Mass, C. (1996). Enhanced green fluorescence by the expression of an *Aequorea Victoria* green fluorescent protein mutant in mono- and dichotyledonous plant cells. *Proc. Natl. Acad. Sci. USA* 93:5888-5893.

Roessner, U., Luedemann, A., Brust, D., Fiehn, O., Linke, T., Willmitzer, L., and Fernie, A.R. (2001). Metabolic profiling allows comprehensive phenotyping of genetically or environmentally modified plant systems. *Plant Cell* 13:11-29.

Royal Society (2000). *Transgenic Plants and World Agriculture*. Document 08/00. London, United Kingdom, <http://www.royalsoc.ac.uk>.

Royal Society (2002). *Genetically Modified Plants for Food Use and Human Health—An Update*. Policy Document 4/02, February.

Schubert, D. (2002). A different perspective on GM food. *Nature Biotechnol.* 20: 969.

Serageldin, I. (1999). Biotechnology and food security in the 21st century. *Science* 285:387-389.

Siemering, K.R., Golbik, R., Server, R., and Haseloff, J. (1996). Mutations that suppress the thermosensitivity of green fluorescent protein. *Curr. Biol.* 6:1653-1663.

Smyth, S., Khachatourians, G.G., and Phillips, P.W.B. (2002). Liabilities and economics of transgenic crops. *Nature Biotechnol.* 20:537-541.

Snow, A.A. (2002). Transgenic crops—Why gene flow matters. *Nature Biotechnol.* 20:542.

Steinberg, T.H., Haugland, R.P., and Singer, V.L. (1996). Application of SYPRO orange and SYPRO red protein gel stains. *Anal. Biochem.* 239:238-245.

Stewart, C.N. Jr. (1996). Monitoring transgenic plants using *in vivo* markers. *Nature Biotechnol.* 14:682.

Stewart, C.N. Jr., Richards, H.A.I.V., and Halfhill, M.D. (2000). Transgenic plants and biosafety: Science, misconception, and public perception. *BioTechniques* 29:832-843.

Tolstikov, W. and Fiehn, O. (2002). Analysis of highly polar compunds of plant origin: Combination of hydrophilic interaction, chromatography and electron spray ion trap mass spectrometry. *Anal. Biochem.* 301:278-307.

Visser, B., Eaton, D., Louwaars, N., and van der Meer, I. (2001). *Potential Impact of Genetic Use Restriction Technologies (GURTs) on Agrobiodiversity and Agricultural Production Systems.* FAO, Rome, April.

Wilm, M. and Mann, M. (1996). Analytical properties of the nanoelectrospray ion source. *Anal. Chem.* 68:1-8.

Zhou, H., Ranish, J.A., Watts, J.D., and Aebersold, P. (2002). Quantitative proteome analysis by solid-phase isotope tagging and mass spectrometry. *Nat. Biotechnol.* 20:512-515.

Index

T - #0484 - 101024 - C0 - 229/152/18 - PB - 9781560222743 - Gloss Lamination